# SOIL CONSERVATION

# SOIL CONSERVATION

*by*

*Helmut Kohnke*
*Soil Scientist*
*Purdue Universtiy*

*Anson R. Bertrand*
*Associate Professor of Agronomy*
*Purdue Universtiy*

BIOTECH

**2014**

**Biotech Books**
**Delhi - 110 035**

ISBN10 81-7622-201-1
ISBN13 978-81-7622-201-3

*Published by* : **BIOTECH BOOKS**
1123/74, Tri Nagar,
DELHI – 110 035
Phone: 27383999
e-mail: biotechbooks@yahoo.co.in

*Showroom* : 4762-63/23, Ansari Road, Darya Ganj,
NEW DELHI - 110 002
Phone: 23245578, 23244987

*Printed at* : **Chawla Offset Printers**
New Delhi – 110 052

**PRINTED IN INDIA**

# *Preface*

The days of agricultural expansion are over. Very little potential farm land remains unused. Insufficient rainfall, prevalence of diseases of man and animals, and remoteness from centers of population are the main reasons that some land is still only slightly used. But by and large, the frontiers have been reached.

At the same time, the population of the earth increases by more than 30 million a year. If we are to feed and clothe everybody, it is clear that the productivity of the land must be increased, not just maintained. Temporary or local food surpluses should not blind us from seeing the over-all problem with all its dire implications.

Agricultural technology has made great strides during the last century. In recent decades the development has been almost explosive. But most of these improvements have been directed toward quick returns. The maintenance of the body of the soil is not so glamorous as the doubling of the yield. For this reason soil conservation has not always received the attention it deserves.

It is the purpose of this book to provide a clear picture of the fundamental nature and cause of soil erosion, of the aims of soil conservation, and of the methods of saving the soil and of maintaining its productivity. It is not the purpose of the book to give recipes for soil management, but to help the reader to understand the soil and what happens to it under various forms of land use, so that he himself can work out the solution of any soil-conservation problem.

This book has been written primarily as a college text. We hope, however, that it will be useful to anyone active in soil-conservation work. While basing our statements on many sources, we have attempted to give a unified picture of the subject.

Soil conservation is more than a technical discipline. It is an attitude. It is the authors' sincere desire that this book may spread the spirit of soil conservation and that it will help to develop a consciousness of stewardship of the soil in the reader.

Assistance in the preparation of this book has come from many sides, especially from numerous colleagues who have shared generously their knowledge and experience with us. We are deeply indebted to all of them. The authors sincerely appreciate outstanding contributions made by Dr. D. L. Allen, Department of Forestry and Conservation, Purdue University; Dr. W. S. Chepil, U.S. Department of Agriculture and Kansas State College; Mr. L. L. Harrold, U.S. Department of Agriculture, Northern Appalachian Watershed Experiment Station, Coshocton, Ohio; Dr. K. Lehotsky, Department of Forestry, South Carolina State College; Dr. L. S. Robertson, Department of Agricultural Economics, Purdue University; and Mr. W. H. Wischmeier, U.S. Department of Agriculture, Agricultural Research Service, Purdue University. Special recognition is due the wife of the senior author, Gerda Kohnke, who has served as critic, consultant, and proofreader.

*Helmut Kohnke*
*Anson R. Bertrand*

# *Contents*

*Soil Conservation*

# 1

# *Soil Conservation as a Problem of Humanity*

## SOIL AS A BASIC RESOURCE

Soil is the medium that makes rain, heat, and light work together to allow plants to grow. The products of plants are many; in addition to food they furnish fiber, fuel, lumber, dyes, and medicines. There are other sources of food besides that which is grown on soil. Fish supply a large proportion of the diet of some people who live near the sea. Of late, plants grown in water cultures and yeast grown in large vats can produce food without soil. But it seems that neither these nor any other artificial method will furnish mankind with more than a small fraction of the required food. To protect our main source of food, the only sensible thing to do is to protect the productive land we now have. The care of the soil is a prerequisite to survival.

Many of the natural resources we find on earth are essential to the life and enjoyment of man. Soil, water, air, energy of the sun, and plants are equally important. Gold, silver, coal, petroleum, iron, and copper are surely valuable, but one would imagine that they could be replaced by other materials. Even livestock is not an absolute necessity in the life of man. Some religious sects, for instance, refuse to eat meat.

While there is no difference in the importance of soil, water, air, heat, light, and plants, there is a definite difference in the destructibility of these resources. For example, water and air circulate. After the water and the air that serve us leave, other quantities replace them. These replacements are not always "punctual," however; there may be a temporary lack of water or air.

After the winter and the night are over, heat and light return. As long as the plant species is not extinct, we can put it into our service.

The story is different with soil. Once the soil is gone from a field, it takes more time than we have for new soil to be developed. For all practical purposes we can say that soil is an irreplaceable resource. Soil is by far the most important irreplaceable resource.

## *Soil Deterioration and Soil Erosion*

**Types of Soil Deterioration.** The dangers that beset the soil are manifold. The most obvious one is soil erosion, the bodily removal of part or all of the soil. But there are other ways in which soil can lose its productive capacity. Nutrient depletion, lack of water, lack of air, excessive physical resistance to root penetration, high salt concentration, and the presence of toxic substances can cut down the yield of crops.

In a virgin soil plant nutrients are usually maintained at approximately the same level over a long period of time. As some chemicals come into solution from the minerals, others are leached out or carried away in the runoff water. Thus a balance is maintained. This may be at a high or low level of productivity, depending on the parent material, the climate, and the topography.

When crops are grown, much of the stock of plant nutrients is removed from the soil by the crops; moreover, tillage operations help to decrease the amount of organic matter. Decrease in the amount of plant nutrients available to crops is brought about by (1) removal of soil; (2) removal of plant nutrients by crops, runoff, and leaching; and (3) decrease in the availability of plant nutrients.

There is an interrelationship between plant-nutrient status, organic matter, and structure of the soil, and a further interrelationship between these and erosion. The productivity of the soil may deteriorate through changes in any of these factors. It is not easy to keep these factors clearly apart.

The long-time trend of yield is not exclusively a result of change of soil quality. Changes in cultural treatments and improvement of plants by breeding have been the major contributing factors to yield trends.

**Extent of Erosion.** The degree to which erosion has occurred differs from place to place. It is sometimes difficult to determine just how much of the soil has eroded. For this reason it is practically impossible to arrive at accurate conclusions concerning the total loss of soil from certain areas. Nevertheless, it is probably fairly correct to say that an average of approximately three inches of surface soil have been lost in the Eastern United States since they have been settled by the white man. The original depth of the "topsoil" of 8 in. has dwindled to 5 in. Even though this figure may be only a rough estimate, it shows that a large percentage of the most fertile part of our soils has been lost and that we should do all that is in our power to diminish this rate in order to maintain the productivity of the soil. In 1934, just before the creation of the U.S. Soil Conservation Service, a survey of the erosion situation in the entire United States was undertaken. From this survey it was concluded that 50 million acres was so severely eroded that productivity was seriously impaired and that another 100 million acres of cropland was on the verge of the same condition. It was also estimated that annually 500,000 acres is rendered unfit for further immediate practical cultivation. Lipman and Conibaere [11] calculated that erosion removes from the land about 20 times as much plant food as crops do.

All these figures have to be used with a great deal of caution, not only because of the difficulty of such surveys, but also because of the evaluation of the seriousness of erosion. Some of the "ruined" land has been put into successful production by the use of lime, fertilizer, manure, sound management, and possibly by a shift to a different type of agriculture.

But while the exact quantities of erosion are not known, the seriousness of the problem is evident.

## *Philosophy of Soil Conservation*

**Definition of Soil Conservation.** According to the dictionary, *conservation* is the act of preserving from decay, loss, or injury. This definition does not include the use of the object that is conserved. Soil conservation would therefore be the preservation of the soil without regard to the use of it. This, however, is not the accepted definition.

Soil conservation has become a widely used concept in the United States only since the importance of the erosion induced by the activities of man has been generally recognized. It is for this reason that we associate and frequently even equate soil conservation with soil erosion control, but soil conservation is more than that. It is not merely plugging up gullies or using more grass or terracing or strip cropping. It is the integration of all these things with sound land use and treatment.

*Soil conservation* is the wise use of land, especially with respect to erosion control. *Conservation farming* is permanent agriculture. It is the proper use of every acre; it is good, sound agriculture with a view to the future. It is maintenance, and sometimes restoration, of soil productivity. It is harmony between man and the land.

Conservation farming means making the most efficient use of the land over a long period of time. Exhaustive farming may yield more cash than conservation farming, but for a short period of time only.

**Need for Soil Conservation.** It is our task to conserve the productive capacity of our soils. It is impossible to restore it. The wealth and the culture of a country depend upon its topsoil. Once this soil is gone, no agricultural miracle will bring it back to full production. Improved methods may increase the productive capacity of a worn-out soil, but the same methods would have resulted in much larger yields if the soils had not been impaired in the first place. Such soil restoration becomes necessary in many cases, but our aim should be the conservation and increase of soil productivity. The term "soil conservation" as we use it today includes all these concepts.

**Relation of Soil Conservation to Other Subjects.** As soil conservation is the management of the soil with the purpose of producing high yields and at the same time protecting it from erosion, a wide variety of subjects is closely related to it. Some of these are soil physics, soil fertility, soil chemistry, soil biology, soil classification, hydrology, climatology, field-crop and pasture management, forestry, agricultural engineering, animal husbandry, agricultural economics, sociology, and wildlife management. To be able to practice soil conservation intelligently, one must acquaint himself with the fundamental facts in each of these subjects.

## SOIL CONSERVATION AND HUMAN WELFARE

### *Population Pressure and Food Supply*

**Area Required to Feed People.** The area of land required to feed mankind depends upon the number of people, the calories required per person, and the calories produced per acre. It also depends on the amount of sea food available.

To express the area required to feed people in acres per inhabitant fails to account for the productivity of the land, the quality of the climate, and the type of agricultural methods used. It might be more appropriately expressed as calorie potential per inhabitant per year.

The area needed to provide food for one person depends upon a multitude of factors. Among these are:

The productivity of the soil and the climate
The intensity of agriculture
The type of crop or product produced
The methods of harvesting, transporting and storing
Direct or indirect consumption
The refining of the products
Eating habits and the preparation of the food
The specific food requirements of the person

All these factors vary greatly from case to case.

The productivity of prairie land in an arid region may be so low that many acres are needed to furnish the food for one person, while on rich tropical land that has an ample supply of water a dozen people can be fed from the products of 1 acre.

The level of production from the same acre depends upon the intensity of agriculture. Improved tillage methods and crop varieties, fertilization, pest control, drainage, and irrigation are examples of practices that can influence greatly the amount of crop production.

The type of crop or product depends largely upon the soil and climate. Several products are compared in Table 1-1.

The degree of completeness of harvest varies with the methods used. Perhaps the outstanding example of this is the cutting of forage as opposed to pasturing it by livestock. While the latter method is frequently the more economical one, it wastes much of the food value by trampling. Appreciable parts of the corn yield are left in the field because corn pickers do not function properly or because the corn stalks are broken or become lodged in the soil. Much of the food value of harvested crops is lost in transportation and storage. Modern refrigeration and drying techniques are helping to overcome some of these losses. It is difficult, however, to preserve all excess products for use in future years.

Aside from the productivity of soil and climate the most significant factor in determining the amount of food per acre that is available for human consumption is whether the agricultural products are used directly by man or are first consumed by animals. Only a small portion of the energy in the feed given to animals is available for human use

*Table 1-1. Human Food from an Acre of Staple Farm Products*

| | Human food produced per acre per year | | |
|---|---|---|---|
| | Weight of food, lb | Digestible protein, lb | Energy, million cal |
| Dairy products: | | | |
| Milk | 3,750 | 125 | 1.22 |
| Cheese | 375 | 97 | 0.73 |
| Butter | 170 | 2 | 0.61 |
| Meat: | | | |
| Pork, 600 lb live weight | 470 | 39 | 1.15 |
| Poultry, 475 lb live weight | 290 | 56 | 0.31 |
| Mutton, 415 lb live weight | 190 | 25 | 0.23 |
| Beef, 370 lb live weight | 215 | 32 | 0.22 |
| Food crops: | | | |
| Corn, 60 bu | 3,350 | 300 | 5.34 |
| Wheat, 34 bu | 2,050 | 190 | 3.06 |
| Soybeans, 27 bu | 1,650 | 500 | 2.62 |
| Oats, 60 bu | 1,350 | 150 | 2.15 |
| Irish potatoes | 10,000 | 110 | 3.26 |
| Sweet potatoes | 10,000 | 90 | 4.88 |

Based on M. O. Cooper, and W. J. Spillman, Human food from an acre of staple farm products, *USDA Farmers' Bull.* 877, 1917.

in the form of animal products. Of all the farm animals, dairy cows excel in the production of food for human beings. With a given amount of feed a dairy animal will produce over five times as much energy and about three times as much protein as is contained in beef made from the same feed. Hogs are more efficient producers of energy than poultry or sheep. Under many conditions, however, the land cannot economically produce plants that could be of direct value in the human diet; for instance, hilly pasture and semiarid range fall into this category. Refining of farm products accounts for a large loss in food value. The prime example of this is the removal of the germ and the protective layers of the wheat kernel before it is milled. Even after food enters the kitchen, the percentage actually used depends largely on its preparation and on individual eating habits. The wealthier the people, the smaller the proportion of the food actually eaten. It has been calculated that the amount of food that becomes garbage in a large American city would be enough to feed all the people in a city in Asia containing ½ million inhabitants. This points at the same time to another factor determining the area required to feed one person—his individual requirements. Age, sex, race, physical activity and social and economic background have much to do with this.

**Changes in Food-producing Areas.** The amount of land area that can be used for food production is not static. There is a continuous increase and decrease of these areas.

*Increases in Food-producing Area.* Draining of sw mps, irrigation of previously unproductive dry land, and reclamation of land from the sea probably account for the largest part of land that is brought into production. Clearing of forests, grading mine dumps with power equipment, removal of rocks, and other reclamation practices are of importance in certain areas. All these methods together promise an increase of some 70 million acres of food-producing area in the United States.

*Causes for Decrease in Food-producing Area.* Food-producing land is lost every year because of poor agriculture and industrial, transportation, and residential developments.

Erosion is probably the cause for the greatest loss of productive land. Other agricultural causes are covering flood plains with infertile deposits; choking streams with erosional debris, thus raising the water level of alluvial lands too high; and deflocculation of soil as a consequence of irrigation with alkaline water. Some of our most fertile valley land is used for water-storage reservoirs.

Industry also takes a large toll of land that is capable of producing food. Examples are mining, especially strip mining; space needed for industrial plants; and storage.

Roads, railroads, airports, and pipelines for oil and gas require each year many new square miles of land that are taken out of agricultural or forestry production. Relatively small losses result from the establishment of electric lines of various types.

The expansion of our cities and the urge to live out in the open remove other areas from food production. It is true that in some sections of the United States the new residential districts are established on land that has never produced agricultural crops, such as deserts and mountains, but by and large, the best agricultural land appears most desirable to the potential home builder. Altogether, more than 1 million acres of land are taken from agricultural production in the United States each year.

**Areas Used to Feed Mankind.** Most people have always had to struggle to find enough food to sustain themselves. Table 1-2 illustrates that in some countries of the world the food supply is very low.

*Table 1-2. Population Density and Food Supply of Certain Countries*

| Country | Inhabitants per square mile | Arable land per person acres | Permanent meadows and pastures per person acres | Calorie content of national average food supply per person |
|---|---|---|---|---|
| Argentina | 17 | 3.97 | 15.00 | 2800 |
| Canada | 4 | 6.40 | 3.58 | 3120 |
| United States | 63 | 2.95 | 3.88 | 3090 |
| Sweden | 46 | 1.29 | 0.25 | 2975 |
| West Germany | 535 | 0.44 | 0.28 | 2945 |
| Netherlands | 828 | 0.25 | 0.29 | 2925 |
| Japan | 616 | 0.14 | 0.04 | 2165 |
| India | 297 | 0.10 | 0.06 | 1840 |

Based on *U.N. Yearbook Food and Agr. Statistics*, 1955, vol. 9, part 1, Rome, 1956.

A large proportion of mankind would be continuously on the brink of starvation if it were not for imports from the countries that produce an oversupply. But unfortunately, the countries with the greatest need for imports are not in a position to pay for such imports, because their exports and services to other nations are too small.

While the food situation of a number of nations, especially in the Western Hemisphere, is very favorable at the present time, prospects for the future are uncertain. In the United States the population has doubled in the first half of the twentieth century, while the acreage of food-producing land has not increased. Improved technology has more than kept pace with the increase in population numbers, but it is doubtful that it can do so indefinitely.

In other countries, for instance, Puerto Rico and Southeast Asia, the situation is much more grave [7].

Better agricultural technology and better sanitation are causing a tremendous expansion of population. It has been estimated that if the world population continues to increase at the present rate of 1½ per cent per year, in A.D. 2300 there will be only 5000 square feet of land surface available per inhabitant [22]. This includes mountains, deserts, and Antarctica. It is obvious that adequate nutrition for mankind would be impossible under such circumstances. Since the land supply is very limited, the capacity of increasing food production is relatively small. Only a restriction of population numbers can, in the long run, guarantee everybody adequate food.

**Potential Increase in Food Production through Improved Methods of Agriculture.** After many centuries of a rather static agricultural technology, improvements and entirely new developments have marked the farming methods of the last two centuries. Progress has been made in all fields, but probably the most conspicuous changes have come in machinery and fertilization. The last twenty-five years have seen agricultural technology take such great strides ahead that it seems reasonable to believe that the future will also witness an immense increase in the producing capacity of agriculture. It is futile to venture a prediction as to how fast and how far this progress will go. There is very little doubt, however, that within a century North America can readily produce enough high-quality food to sustain a population twice as large as the present one and do it at a high physical standard of living.

It is obvious that this or higher goals can only be reached if no world catastrophe occurs and if every effort is made to maintain and improve the productivity of all soils.

It should not be overlooked that the northern regions of this continent contain vast areas of potentially productive land that has not been utilized because of the harsh climate and the short growing season. Modern brush-clearing methods and housing techniques as well as a knowledge of fertilization make large-scale settlement and agricultural production in some of this area feasible.

## *Soil Management and Human Health*

**The Sequence: Soil–Plant–Animal–Man.** Plants grow in the soil and derive an essential portion of their nutrients from the soil [4]. Animals eat plants. Man, who eats both animals and plants, therefore depends for his sustenance and well-being on the soil as well [16]. If the soil is deficient in a certain element, the plants will be deficient in the same element, and consequently also animals and man, who eat these plants, will suffer [2, 18]. People in technologically advanced areas use food from a multitude of sources and therefore seldom suffer from lack of a specific element. Also, extensive use of sea food overcomes detrimental effects that might result from mineral deficiencies of a diet derived entirely from the land.

Some of these deficiencies are geographically widespread and consequently affect large populations. Calcium and phosphorus are the elements which occur in inadequate amounts in many parts of the world, particularly in the sandy and the highly leached soils.

The two other elements that are likely to be deficient in our diets are iodine and iron. If vegetables, milk, and meat do not contain the proper minerals and vitamins in sufficient quantities, then the human body cannot be maintained at maximum health and vigor.

**Soil Erosion and Human Health.** Erosion may have a detrimental effect on human health in two ways: In many soils, especially the leached forest soils, calcium, phosphorus, and other minerals are concentrated in the upper layers of the soil. When these are removed by erosion, plants grow on the more deficient subsoils and therefore may contain less of these minerals.

The second reason is that generally crop yields are smaller on eroded soils than on soils with a complete profile. This may mean that the farmer receives too small an amount of food from his land and so small an income that he cannot afford to purchase adequate supplementary nourishment.

It should be kept in mind that plants are reduced in growth rate and eventual size if the soil is "poor," either inherently or through erosion. But the composition of these plants with respect to minerals may be the same as, or even superior to, plants that have grown on "rich" soil. Plants grown on the latter are frequently relatively high in nitrogen and carbohydrates but low in minerals. We may conclude from this that erosion has a detrimental effect on the composition of human food only in a minority of cases, but it can affect the health of local people through the poverty it creates.

**Food Requirements and Eating Habits.** In considering the area needed to produce the food required to feed mankind it has to be kept in mind that food requirements per person vary greatly with eating habits. These are determined by a variety of causes.

*Climate.* More food is needed in cold than in warm climates. The colder the climate, the greater also is the need for fat.

*Housing and Occupation.* Modern housing and a sedentary occupation reduce food requirements. In fact the effect of climate on the food requirement of a person who spends most of his time in air-conditioned or heated rooms is negligible.

*Availability of Food Types.* Climate, soil, and customs determine the types of food produced in a given area. Because of their accessibility and their relatively low prices they usually are the favorite foods of the inhabitants.

*Economic Conditions.* The general economic conditions, the specific economic status of the individual, and food prices compared to one another and to other goods determine to a large extent the quality and quantity of food consumed. At present, the cost of food in the United States is about 20 per cent of the average income. What would happen if it were to rise to 50 per cent?

*Technology.* Technological developments along the lines of producing, refining, and storing of food have much to do with eating habits. Sterilizing and freezing of otherwise perishable food has made most foods available anytime almost anywhere. The refrigerator has greatly changed eating habits in the last half century, especially in countries with hot summers. The various methods of preserving food are helping to make better use of the available supplies. Another technological development is the refining of foods. While generally desirable, refining has the tendency to remove large portions of the nutritionally essential elements. The complete removal of all minerals and vitamins in the refining of sugar is only one example.

**Commercial Fertilizers and Food Quality.** The literature of the last decades contains many contradictory statements concerning the effects of fertilization on food quality. There is a group that believes that chemical fertilizers, per se, are detrimental to the nutritional value of crop plants. This is based on the assumption that such fertilizers are "unnatural" and therefore must impair the plants. Others are just about as certain that poor soils produce poor-quality foods and that fertilization is bound to be beneficial [1].

Actually fertilization may improve or impair crop quality. If a soil is low in a specific element, say phosphorus, addition of available phosphate is going to improve the nutritional value of the crops grown on it. If, on the other hand, the same phosphorus-deficient soil is "fertilized" with nitrogen, the ratio of the various nutrient elements is thrown out of balance still further and naturally the value of the plants grown as feed and food is greatly reduced. Whether fertilization will be beneficial or detrimental to the nutritional value of the crops grown depends entirely upon the resultant combination of plant nutrients. If the addition of fertilizers makes this combination more favorable for plant growth and health, its use is also of advantage to the animals or the people that consume the plants. And it is of no consequence whatsoever whether "artificial" fertilizer or organic substances are used to bring about this condition.

Nutrients taken up by plants from the soil are in ionic form. Therefore the plant foods in organic materials must be reduced to these forms, which are essentially the same as in the compounds in commercial fertilizers.

It should not be overlooked in this connection that the amount of crops required to feed mankind satisfactorily is so large that it would be completely impossible to produce that amount if it were not for the extensive use of commercial fertilizer. And the amounts of fertilizer used must of necessity increase as the world population increases.

### *Rural Living Conditions*

If it is our aim that soils be maintained in a state of permanent high productivity, it is necessary that living conditions on farms be attractive enough to hold a sufficient number of well-qualified, healthy people. Throughout history living conditions have frequently been poorer in the country than in the cities. While country life has many advantages over city life, the longer working hours and the lack of many of the conveniences that we like to call "modern" have sent an unending stream of farmers' sons and daughters into urban areas. This relationship between living conditions in the city and on the farm has changed from time to time and from place to place. In the United States rural living conditions have improved so much that the incentive to leave the farm has greatly decreased.

Let us compare the individual items that determine living conditions on the farms with those in the cities in the United States.

*Nutrition.* Nutrition on the farm is generally satisfactory and frequently even better than in town. Where much of the food is grown on the farm, the danger of mineral-deficient nutrition does exist in some areas if little outside food is bought. Canned goods, refrigerators, and deep freezers are a great help in permitting the use of a variety of food of high quality.

Drinking water, a very essential ingredient of the diet, is not always what would be desired on the farm. This is particularly true in semi-arid regions.

*Housing.* Today there is very little difference in housing between town and country. Most rural residences have electricity, and many electric appliances are used. Running water and central-heating systems have become standard.

*Working Conditions.* Mechanization has made farm work less burdensome and more attractive. It has taken away much of the drudgery formerly connected with it. Nevertheless, most farmers still have to work longer hours than their city cousins.

*Transportation.* Perhaps the greatest change in living conditions on the farm has been brought about by the introduction of the automobile. Most farmers in the United States can now get to town in less than an hour. "Rural free delivery" by automobile brings them their mail daily.

*Education.* In spite of school buses and consolidated schools, education of the farmers' children is frequently not as thorough as that of city children. Schools are smaller, facilities not adequate, and most teachers prefer city life. Great improvements have taken place in country schools during the last fifty years, and the prospects for the future are bright. To compensate for some possible inadequacies of rural schools many organizations have made it their task to carry information to both youth and adults in the country. Some examples are "4-H," Future Farmers of America, and the extension services of state and Federal organizations which include county agents and home-demonstration agents as well as the agricultural advisors of various industrial enterprises.

*Medical Care.* In the densely settled parts of the country there is practically no difference between the quality of medical care in the country and in the city, since the motor car helps to bring patient and physician together without much delay. In the more remote sections of the country this is not the case. Many country doctors still have to care for an unbelievably large area.

*Churches.* Farmers usually find a church of their choice within a reasonable distance from their homes.

*Entertainment and Recreation.* Radio, television, improved transportation, and many features of modern entertainment and recreation do much to make country life more attractive than it used to be.

*Income.* To attract qualified people to the country the incomes of the farm owner, the tenant, and the hired man should compare favorably with those of people of similar responsibilities in town. Whether this is the case or not depends on the current economic situation. The hired man, whose wages until the late 1930s were about the lowest of any laborer's, has seen his income climb steadily, in line with his greater effectiveness due to mechanization.

There are many items that determine the level of living conditions of the farmer and his family, such as his own intelligence and aggressiveness, the size of the farm, the quality of the soil, and the market conditions. Consequently, there will always be a wide range of living conditions, as there is also in town. This is as it should be. The only thing in which the commonwealth as a whole should be interested is to make certain that farming conditions are made attractive to the best of the population. Farming is becoming increasingly complex and requires high-caliber men. If farming is made pleasant, there will not be a flight of the best talent to the cities.

## GEOGRAPHY AND HISTORY OF SOIL EROSION

### *Correlation between Geography and History of Man-induced Erosion*

It is difficult to separate a discussion of the geography of man-induced erosion from that of its history. These two are closely interrelated. Climate and soil characteristics are largely responsible for the density of population of a certain region. The time of occupation and the form of land use by man determine the amount of erosion—other things being equal. The areas that were adapted to the agriculture and the habitation of the epoch were the ones that were settled first, and it is these that first suffered from man-induced erosion. This is probably the reason that much of the land in the Near East, where a temperately warm climate prevails, was densely populated at the dawn of history and has long lost most of its productivity through erosion [12].

On the other hand, areas that are difficult to reach because of climatic extremes or are undesirable because of the prevalence of diseases, such as the subarctic regions or the Amazon Valley, have retained their soil cover. The early occupation of the East and Southeast of North America by energetic people is chiefly responsible for erosion being much more severe there than in other areas of the continent that have been settled later, although climatic differences are also involved.

Practically all phenomena that affect the degree of erosion are geographically determined. They include climate, topography, soil characteristics, productive capacity for certain crops, population density, as well as the economic, industrial, and political situation of the area.

## *Areas of Severe Erosion*

Where the various causes for erosion have existed for a sufficiently long time, much of the original surface soil has disappeared from the land. In addition to the causes already mentioned, the type of land management is probably the outstanding reason why soils are badly eroded in one area while in another they still retain much of their original profiles.

Only a few examples can be given here. The loess region of China is so eroded that the student of soil genesis finds nearly complete profiles only in cemeteries, where the ground has been protected for many centuries. Loess is very erodible, and in China it has been cultivated for about four thousand years. Only where terraces for rice paddies have been constructed has much of the soil been conserved.

The Near East, from the Mediterranean to India, is another area that has suffered much from soil erosion. Climate and geographic location caused this area to be settled early. Dense populations have existed here through thousands of years, and at times the food requirements were so high that exhaustive agriculture was practiced. Uncontrolled pasturing with sheep and goats is practiced in many of these areas to this day. Thus vegetation has been nearly eliminated, exposing the soil to the ravages of wind and water. As a result large stretches of the Near East are very unproductive or even actual deserts today.

Large parts of the area that is today the Sahara Desert maintained a heavy population and an intensive agriculture a thousand years ago [3, 8]. The warm climate of the region has not been conducive to the accumulation of organic matter in the soil, and here, too intensive land use has caused the complete destruction of the agricultural potential.

The United States is a prime example of how intensive agriculture, particularly monoculture, has been detrimental to the fertility of soil where climate and topography have been conducive to erosion. Much of the steeper land in the Southeast has lost all of its topsoil after it had been cropped to cotton for a century or two [9]. Severe rains and low organic-matter content of the soil have combined with exhaustive agriculture and sloping topography to cause serious sheet erosion and probably the largest gullies existing in the country.

In the case of the Dust Bowl that includes areas of Kansas, Colorado, Oklahoma, Texas, New Mexico, Arizona, and Utah, the cause of severe erosion has been continuous cropping to wheat and overgrazing with sheep. The rainfall in this area is sufficient for fair crop growth for a period of several years. This encourages farmers to fairly intensive use of the land. In the dry part of the weather cycle that invariably follows, the strong winds that occur in the Great Plains sweep unobstructed across a soil that has lost much of what little organic matter it possessed. The result is that the soil cannot hold together and is blown into the air, causing health hazards to the inhabitants and economic ruin to the farmers. Since this area has been cultivated for less than a century, the accumulated effects of erosion have not been so severe—except locally—that intelligent, adapted land use could not restore it to productivity.

One of the areas of most serious erosion in the Dust Bowl is the Navajo Indian reservation in northwestern New Mexico. Both overgrazing with sheep and a climate that combines hot dry winds and occasional torrential downpours have caused severe wind and water erosion. The carrying capacity of this land has greatly diminished. Much of it is not much better than a desert.

### *Areas of Insignificant Erosion*

To bring about serious man-induced erosion, soil must be exposed without vegetative or other protection to a climate with either heavy rains or strong winds. Obviously there are many areas on earth where this combination does not exist.

**Areas without Agriculture.** Land may be too stony, too steep, too swampy, or too cold to be used for farming. On such land grazing is usually too light to destroy the vegetative cover. Examples of such

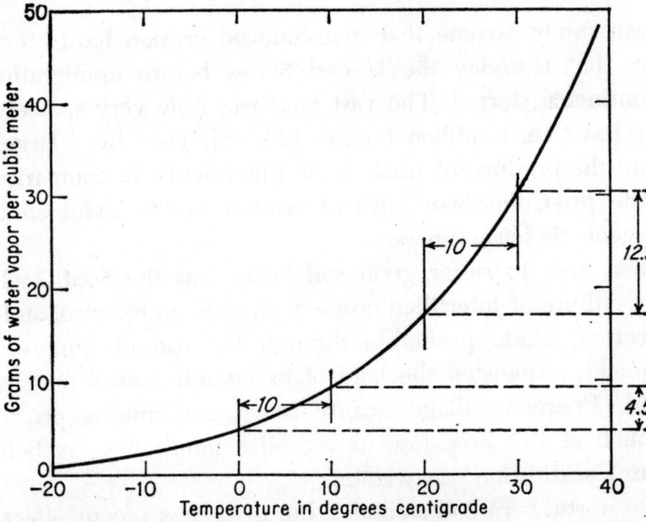


**Fig. 1-1.** Water vapor in saturated air. This curve shows that the amount of water vapor in saturated air increases greatly with temperature. A drop of temperature from 10 to 0°C condenses only about one-third as much water vapor as a drop from 30 to 20°C.

areas are the forests in mountains and on hilly land and the only partly explored vast regions of Central Africa and the Amazon Valley. There is also practically no agriculture in the northern half of Canada and in much of Siberia. Pasture land that is not overgrazed falls also in this group.

**Areas with Gentle Rainfall.** Severe rainfall is usually the result of a large temperature drop in areas of warm, moisture-laden air. Cold air, even if saturated with water vapor, contains only a small amount of water. Further cooling can cause only slight precipitation. Therefore rainfall in cool climates is of low intensity, and moreover, much of it comes down in the form of snow. In certain other areas there is much moisture in the air, but the lack of cold air masses or of mountains to cool down the moist air is the reason that rainfall is only gentle. England is a prime example of this condition. In spite of over a thousand years of cultivation there is little evidence of severe erosion. Much of Northern Europe is in a similar condition.

### *Erosion History of the United States*

We can safely assume that man-induced erosion hardly existed in the area that is today the United States before immigration from other continents started. The vast land was only very sparsely populated by less than a million Indians [15, 21]. They lived largely from game and the produce of small, hand-tilled fields. In comparison with today, the pre-Columbian rates of erosion can be assumed to have been of geologic leisureliness.

The first area to suffer great soil losses was the South, where an intensive culture of intertilled crops, such as corn, tobacco, and particularly cotton, made profitable through the use of imported slave labor, quickly exhausted the land of its organic matter and its nutrients [13]. Frequent tillage, scanty cover, and intense precipitation made much of this area lose its topsoil a hundred years before the American Constitution was written.

Farther north, corn had a similar, but much less severe, effect on the soil. East of the Appalachians erosion was already a problem in colonial days. The longer winters and the somewhat smaller amount of rainfall, together with the higher content of soil organic matter—compared to the conditions in the South—retarded the damage. As the farmers moved farther west, erosion followed. Many farmers reaped several years of good crops, and when the yields decreased because of soil erosion or nutrient depletion, they went on to new land to repeat the process.

Most of the early immigrants to North America came from Northern Europe, where climatic conditions are much more gentle and erosion is hardly a problem. It is natural that they did not realize its dangers when they started to farm on American soil. To most of the early farmers the soil was a resource that could most profitably be used by making quick use of its productivity and abandoning it when it was exhausted. This is an entirely understandable policy if we consider that commercial fertilizer did not yet exist and seemingly endless areas of fertile land could be found in "the West."

There were a number of exceptions to this pioneering farming sequence of exploration–exploitation–exhaustion. The best known of them are the Pennsylvania Germans, who cleared the sloping land of Lancaster County three centuries ago. Many of the farms are still in the same families, and the land is as fertile as ever. These people came from the hilly sections of western and southern Germany, where erosion was a well-known hazard of farming and where they had learned the value of rotations and of contour cultivation.

The introduction of power equipment to till the soil has introduced a new great danger for American soils. Properly used, power implements can help to conserve the soil, but on the other hand, they give the farmer the chance to bring large areas under cultivation, thus exposing more land to wind and the impact of rain. Their great weight also helps to compact the soil and diminish its infiltration capacity, thus enlarging runoff and erosion.

Much erosion has occurred in the semiarid areas of the United States. While wind erosion is more spectacular, the greatest damage is due to water erosion. Dryness is the cause for a low organic-matter content of the soil. Consequently, the soil is poorly aggregated and washes away when heavy downpours occur. These are not as frequent as in the Middle West and the South, but they are frequent enough to wash away much of the topsoil. It is not easy to distinguish between geologic erosion and man-induced erosion in semiarid regions. The Spanish explorers gave the largest river of the semiarid Southwest the name "Rio Colorado," the colored river, because of the large quantities of red soil it carried. That was at a time when practically no agriculture existed in its watershed.*

Erosion continues in practically all parts of the United States. But since the early 1930s the realization that soil erosion is a national menace has become general among the leading citizens in country and city alike. Public and private organizations have been created both to combat erosion and to publicize its importance. While no accurate estimates exist, it seems certain that the rate of erosion—over the country as a whole—has already diminished because of the many con-

* The word *watershed* frequently denotes two different things:

1. A drainage area, e.g., the Missouri Basin equals the Missouri Watershed
2. The boundary of a drainage area, e.g., the Continental Divide

In this book "watershed" is used to mean drainage area.

servation features that have been introduced into American agriculture.

## SOIL-CONSERVATION ORGANIZATIONS

Soil erosion has been recognized to be a problem of such far-reaching importance that its control cannot be left exclusively to the resources of the individual farmer. While much of the actual work is done by him, he requires advice and technical assistance. Some soil-conservation practices are feasible only when they are carried out over areas that are much larger than a farm. Examples are wind-erosion control, extensive drainage projects, and runoff in stream valleys. Also, flood-control dams can be built only by community effort.

The entire population, both rural and urban, depends on agricultural products, and the farmer is an important unit of every nation. It is therefore the responsibility of the people as a whole to lend the farmer assistance in conserving the soil. Assistance is needed in the survey of the existing erosion condition, in research into methods of erosion control, in advising the best methods for specific cases, in education, in financing, and in the administration of large-scale conservation projects.

A variety of organizations has been created throughout the world to perform these functions. They are of local, state or provincial, national, international, or private origin. Some of these organizations are created specifically for the conservation of soils, while the majority devote only part of their efforts to this task.

The Food and Agriculture Organization of the United Nations contributes to soil conservation both by making soil-erosion surveys and by sending soil-conservation specialists to areas where there is a lack of trained personnel.

Probably the most elaborate set of soil-conservation organizations exists in the United States. The main reason for this is that here the change from the primeval condition of complete soil protection by forests and prairie grasses to severe man-induced erosion was faster than in any other large country. While strip cropping was practiced by the early Pennsylvania Germans and terracing has been done in the Southeast since colonial times, it took more than a century from the first recorded recognition of soil erosion to the creation of an agency devoted to its control [17]. While soil-conservation research has been conducted in the Middle West since 1917 by some of the land-grant colleges [14] and since 1928 on a national scope, the first organization devoted specifically for the control of erosion was the Soil Erosion Service. It was created in 1933 as a temporary agency of the U.S. Department of the Interior. The great dust storms in the West and the floods in the East that occurred in the middle thirties

**Fig. 1-2.** Soil-conservation district in action. A soil surveyer discusses the significance of a soil profile on land-use planning with a soil-conservation district board in Illinois. (*USDA*)

helped to focus the attention of the nation on the problem of the interrelation between land use and soil condition. Upon the urging of H. H. Bennett, the Soil Conservation Service was established in 1935 as a permanent "bureau" of the U.S. Department of Agriculture. It continued the demonstrational work that had been started by the Soil Erosion Service. It also took over soil-conservation research from other bureaus and expanded it considerably. During the 1930s considerable assistance was given the Soil Conservation Service by the Civilian Conservation Corps, an organization created with the double purpose of supplying labor to carry out many of the soil-conservation tasks, such as planting trees, building gully-control structures, and improving road banks and of providing useful work for young men during the Depression period. The Works Progress Administration followed similar aims for adult men; its program, however, was of a more inclusive nature. Beginning in 1937, the Soil Conservation Service has put its main emphasis on cooperation with the "soil-conservation districts" [17]. These are local organizations of farmers

**Fig. 1-3.** Soil-conservation-district meeting. A vocational agriculture teacher helps farmers to understand the importance of soil characteristics in farm planning. (*USDA.*)

whose aim is to help the members in designing and adopting soil-conservation measures. The districts are voted into being by the farmers of a specific area, frequently a county. The officers are elected from the land owners or land occupants of the district. Usually the districts apply for technical assistance from the Soil Conservation Service. Such assistance makes up the majority of the activity of the Soil Conservation Service (Figs. 1-3 and 1-4). It includes soil and erosion surveys, farm planning to reduce erosion losses, and engineering of waterways, terraces, and the like.

In 1954 the United States Congress passed the Watershed Protection and Flood Prevention Act (Public Law 566, Eighty-third Congress) with the aim to encourage soil conservation and flood control on watersheds of intermediate size. Federal funds are made available to local organizations to carry out works of improvement for building of dams and dikes for agricultural water management, including irrigation and drainage, and also for the development and management of municipal and industrial water supplies. It is the main purpose of

Fig. 1-4. Land-judging contest. In many states the land-grant colleges and the Soil Conservation Service team up in helping high school students learn about soil properties. Here, a land-judging contest serves this purpose. (*USDA*)

this act to supplement the other soil- and water-conservation programs of the country and the flood-protection work of the major river valleys. It is designed to bridge the gap between these two types of programs and greatly enhance the ultimate benefits of both.

Other Federal agencies in the United States that work on some phase of soil conservation include the Forest Service, the Agricultural Research Service, the Agricultural Conservation Program Service (formerly the Agricultural Adjustment Administration), the Federal Extension Service, the Bureau of Reclamation, the Farm Credit Administration, and the Corps of Engineers of the United States Army [5].

State and local organizations that are concerned with soil conservation in one way or another are the land-grant colleges, flood-control commissions, stream-pollution boards, departments of conservation, and the conservancy districts. A number of private organizations exist that are created specifically for the interchange and dissemination of knowledge on soil conservation. The outstanding ones are the Soil Conservation Society of America and the Friends of the Land.

Soil-conservation problems in other countries are different from those in the United States; therefore the organizations set up to solve these problems are different. In Western Germany many fields have become too small for successful farming because they have been split up every time they passed from a father to his children. As these tiny parcels of land are redistributed so that the pieces belonging to an individual farmer are large enough for efficient operation, an effort is made by the land-adjustment agency to divide the land and to place the roads in such a way that runoff and erosion are at a minimum. Another government agency in the same country, the Department of Landscape Conservation, is charged with the integration of all efforts to maintain or create a pleasant and healthful landscape, a task that is particularly important in a densely settled area [10]. Practically every country in Europe has enacted legislation and has initiated programs for the purpose of soil and water conservation. The classical work in this respect has been done in France, even though the legal provisions making the conservation of soils mandatory only date to the years 1860 and 1864. The earliest record of conservation work through forestry in France dates to the year 1322, in which the city of Briançon passed an ordinance forbidding the cutting of trees in forests which acted as protection against the silting of certain properties that were under the jurisdiction of the city. Several laws were passed by local magistrates endeavoring to protect their land and buildings by means of soil conservation brought about by the maintenance of forests. These local laws were apparently consolidated and ameliorated in the texts of the laws of 1860 and 1864.

Switzerland, Austria, and Italy also have found it necessary to protect mountain slopes from erosion and valleys from sedimentation and have created government agencies for this purpose.

More recently India, the Union of South Africa, Argentina, Colombia, and many other countries have established organizations to develop soil-conservation and land-use techniques and to help the farmers to adopt these [6, 19, 20, 22].

## REFERENCES

1. Albrecht, William A.: Soil fertility and national health, *J. Am. Soc. Farm Managers and Rural Appraisers,* **8**:45–66, 1944.
2. Auchter, E. C.: The interrelationship of soils and plant, animal and human nutrition, *Science,* **89** (2315):421–427, 1939.
3. Ball, J.: Problems of the Libyan desert, *Geog. J.,* **70**:21–38, 105–128, and 209–224, 1927.
4. Beeson, K. C.: The mineral composition of crops with particular reference to the soils in which they were grown, *USDA Misc. Publ.* 369, 1941.
5. Brown, C. B.: State legislation for watershed and flood prevention, *J. Soil and Water Conservation,* **10**:286–289, 1955.
6. De Castro, F. S.: Experimentos sobre la erosión de los suelos, Federación Nacional de Cafeteros de Colombia, *Bol. Téc.* 6, 1951.
7. Flannery, R. D., and E. E. Foster: Land and water problems of East Pakistan, *J. Soil and Water Conservation,* **10**:89–93, 1954.
8. Gautier, E. F.: "The Sahara: the Great Desert," Trans. by Dorothy Ford Mayhew, Columbia University Press, New York, 1935.
9. Hall, A. R.: Early erosion control practices in Virginia, *USDA Misc. Publ.* 256, 1937.
10. Kragh, Gert: Naturschutz und Landschaftsgestaltung, *Deut. Akad. Landwirtschaftswiss. Berlin, Sitzber.,* **5**:21–27, 1956.
11. Lipman, J. G., and A. B. Conybeare: Preliminary note on the inventory and balance sheet of plant nutrients in the United States, *New Jersey Agr. Expt. Sta. Bull.* 607, 1936.
12. Lowdermilk, W. C.: Conquest of the land through 7,000 years, *USDA Soil Conservation Service Inform. Bull.* 99, 1953.
13. McDonald, Angus: Early American soil conservationists, *USDA Misc. Publ.* 449, 1941.
14. Miller, M. F., and H. H. Krusekopf: The influence of systems of cropping and methods of culture on surface runoff and soil erosion, *Missouri Agr. Expt. Sta. Research Bull.* 77, 1932.
15. Mooney, James: The aboriginal population of America north of Mexico, *Smithsonian Inst. Misc. Collections,* vol. 80, no. 7, 1928.
16. Orr, J. B.: "Minerals in Pastures and Their Relation to Animal Nutrition," H. K. Lewis and Company, Ltd., London, 1929.
17. Partain, L. E.: A brief history of soil conservation districts, *J. Soil and Water Conservation,* **10**:9–12, 1955.
18. Pottenger, F. M., Jr., Ira Allison, and W. A. Albrecht: Brucella infections: Possible relation to deficiency of trace elements in soils, plants, animals and man, *Merck Rept.,* July, 1949.
19. Progress report on soil conservation in the European farming area of the

Union of South Africa (with notes on erosion caused by roads and rainways, uneconomic subdivision of land, and protection of headwater catchments), mimeographed report, 1956, pp. 1045–1055.

20. Quevedo, C. V.: Conservación del suelo, Editorial *Suelo Argentino,* Buenos Aires, 1943.

21. Swanton, J. R.: Indian tribes of North America, *Smithsonian Inst. Am. Ethn. Bull.* 145, 1953.

22. Villard, H. H.: Some notes on population and living levels, *Rev. Econ. and Statistics,* 37:189–196, 1955.

# 2

# The Soil

## THE BASIS OF SOIL STUDY

Soil is that part of the earth's crust that is penetrated by plant roots. It is composed of mineral and organic matter. Chemical, physical, and biological factors contribute to the development of soils. The soil may be likened to a natural, living body, inasmuch as it has an embryonic state, followed by a period of development, and then passes into maturity and old age.

The soil body is actually teeming with plant and animal life which together with the purely chemical and physical forces cause the soil body to become organized into distinct parts or zones. The old concept of the soil as a mixture of mineral and organic material is inadequate. An animal which has passed through a meat grinder is no longer an animal body; it is animal material. A mixture of mineral matter and organic matter is not soil; it is soil material.

Throughout history the recognition of natural units as distinct bodies has been the first step in answering the question, "What is it?" To discover the natural laws that govern the behavior and function of a body, one must have a thorough understanding of the parts of the body. Botanists were pioneers in developing modern methods of morphology in the study of plants. The external and to a certain extent the internal attributes were studied and used to characterize plants. Because of this natural approach the botanists made remarkable progress. Great as were the advances made in botany and zoology solely through the application of our natural senses, these achievements cannot be compared to the advances made when such studies have included scientific experiments and deduction involving the use of chemistry, physics, and mathematics to find the answers.

This approach, so successfully applied to other sciences, was slow in coming to the field of soils. For centuries man took the soil more or less for granted. As his needs increased or as the soil became exhausted and refused to yield, he looked for new land. As long as there was enough land, there was no pressing need for soil studies. In some parts of the world colonial empires resulted in disastrous neglect of the soils of the mother country.

In many areas of the world land taxation has been the chief source of state revenue. The fertility of the soil, its yield capacity, and its adaptability to various crops had to be evaluated to fix tax rates equitably. This led to attempts at land classification. Scientists in many fields became interested in the study of the soil. Our present knowledge of the soil is the sum of the contributions of the ancient philosophers, naturalists, and alchemists, as well as modern physicists, chemists, geologists, microbiologists, and plant physiologists. These studies have led to the development of *pedology*, the science dealing with the laws of origin, formation, and geographic distribution of the soil as a body in nature, and have also led to many answers which shed light on the basic question, "What is the soil?"

## SOIL FORMATION AND THE SOIL PROFILE

The soil body when cut open displays a series of layers, designated as "horizons." Each horizon has certain morphological, chemical, and physical characteristics. The succession of horizons, as seen in the exposed anatomy of the soil, is known as the "soil profile." The profile characteristics found at any given spot of the earth's surface result from the particular combination of chemical, physical, and biological factors present. In cold or arid areas, physical forces of weathering predominate, while warm or humid climates cause rapid chemical changes in soil material.

The factors that determine the types of profile formed are:

1. Climate (temperature, water, wind)
2. Biotic factors (plants and microbes, animals, man)
3. Relief (topography or the slope of the land surface)

4. The parent material (the original rocks)
5. Time

Soil material which has been stationary long enough for local climatic and biotic factors to act upon it will have developed characteristics which are common to all other similar situations. These characteristics will be influenced by the parent material and the slope of the land surface.

In an undisturbed woods or virgin prairie in a temperate humid climate, the pedologist usually finds a soil profile which has three distinct horizons, *A*, *B*, and *C*, each of which may be divided into subhorizons.

*The $A_{00}$ Horizon.* From the surface to a depth of from ½ to several inches, depending on the climate and drainage, there is a deposit of organic matter. Its composition varies with the type of natural vegetation. In a forest it consists of leaves and twigs. In a virgin prairie it is dead blades and stems of grass. The materials composing this layer are only partially decayed and may easily be identified.

*The $A_0$ Horizon.* Under the $A_{00}$ horizon will be a layer of similar nature, but which has undergone considerable decomposition. The color is dark brown or black, and identification of the original organic material is difficult or impossible.

*The $A_1$ and $A_2$ Horizons.* The $A_1$ horizon is the beginning of the mineral portion of the soil body. It is usually dark brown to black. In forested regions the color tends to change in the $A_2$ to a light gray, brown, or yellow, depending on the geographic region. In prairie regions the dark color of the $A_1$ horizon may extend to a considerable depth, and the $A_2$ horizon may not be clearly visible.

*The B Horizon.* This horizon lies below the *A* horizon and is the layer of clay accumulation. The upper limit of the *B* horizon is readily recognizable because its color, texture, and structure differ from the *A* horizon. As a rule the *B* horizon contains very little organic matter. The texture is finer than that of the *A* horizon and therefore more compact. The structural units of the *B* horizon are larger than the units of the *A* horizon. The depth, color, and compactness of the *B* horizon vary with the location.

*The C Horizon.* The *C* horizon is usually lighter in color than the *B* horizon. The *C* horizon is practically unweathered. It represents parent material similar to that from which the *A* and *B* horizons were derived by soil-forming processes.

Extremes of any of the five factors which regulate soil formation will result in the development of incomplete soil profiles. A common example is the *AC* profile on steep slopes where the *B* horizon has not developed. Other examples are the hydromorphic soils, which are formed under conditions of excess moisture and consist of depositional layers. The halomorphic soils are formed under imperfect drainage conditions in arid regions. They are characterized by high salt concentration in the upper layers and do not have fully developed *ABC* profiles. Young soils, such as recent alluvium and shifting sand, will show little or no horizon differentiation in the profile.

## FACTORS AFFECTING SOIL FORMATION

### *Climate*

Major variations in soils over large geographic areas can usually be attributed to changes in climate. Climate affects the formation of soils both directly and indirectly. Climate acts directly through precipitation and temperature as they affect the weathering of the parent material. Indirectly, climate controls the biological environment of the soil by determining the numbers and species of plants which grow. To appreciate fully the role of climate in soil formation one must examine both precipitation and temperature separately.

**Precipitation.** Water which falls on the soil as precipitation does not remain at the spot of impact, but rather it percolates downward or runs off over the surface. Percolating water takes with it soluble materials from upper to lower zones or perhaps even entirely out of the soil body. Solid material, mostly colloidal, is also moved downward and deposited below the surface to form *B* horizons. Thus the horizons of the profile become differentiated. Not all the water moves down in the profile. Runoff, evaporation, and transpiration reduce the amount of water which actually moves into underground flow channels. Evaporation and transpiration from the surface and capillary movement of soil moisture may be very influential in affecting the profile formed. Runoff and the resulting removal of surface soil by erosion affect the abnormal soil characteristics.

In analyzing precipitation as a factor of soil formation one must consider not only the quantity but also the forms of water, such as snow and rain, as well as its seasonal distribution, runoff, and evaporation.

**Temperature.** Different types of soil profiles will develop under the same precipitation in two different temperature belts.

In cold zones, where permanently frozen zones exist in the soil, very little percolation and leaching is observed. Even in a subarctic climate, where severe winters persist, conditions are unfavorable for rapid percolation of water through the profile. Organic matter accumulates at the surface and does not entirely decay. The result is a spongy organic layer which protects the underlying mineral material and slows down weathering.

In the northerly section of the humid temperate zone, the soils are highly leached, and the lower portion of the *A* horizon is sometimes ash gray and is generally lighter in color than the upper portion. As one moves to a more arid climatic zone, the trees give way to grass, and the *A* horizon is dark and high in humus to considerable depths. In semiarid and arid areas where the temperature often soars in summer, evaporation is very rapid, and soil formation is slow. These soils are not leached of soluble salts. The *B* horizon is light in color and contains veins and concretions of carbonates of calcium and magnesium.

In humid regions, the *B* horizon contains no carbonates of lime, and the color is usually brown or reddish brown. The compactness of the *B* horizon increases until a certain depth is reached and then begins to decrease. Sometimes this horizon may have iron or iron-humus concretions or even an entire layer of these.

Because of the favorable temperature and moisture relationships in the humid tropics, there is luxuriant vegetation. In soils which are not swampy the optimum conditions for decomposition result in very little accumulation of organic matter. In swamp areas peat formations may develop.

### *Biotic Factors*

The biotic environment of the soil consists of several important groups of organisms, namely, microorganisms, vegetation, animals, and man.

Very little information is available concerning the distribution of microorganisms in various soil types, but what is known indicates that each soil possesses its own characteristic microbial population. Many experiments have shown that changes in the composition of a soil are accompanied by changes in the microbial population. A soil which has a low population of a particular species of microorganisms may abound with this organism very soon after an alteration such as neutralization or aeration.

Because of the relative ease of transport of many of the microorganisms, most soils have been inoculated or reinoculated many times by the majority of soil microorganisms. We may therefore assume that the microbiological factor of soil formation is potentially the same for a given geographic area.

It has long been known that plants are by far the most influential biotic factor in soil formation. Each plant association exerts a specific effect on the soil-forming processes. As plant communities develop from the initial introduction to the climax associations, the soil develops also. The soil beneath the climax vegetation often has characteristics of a mature profile.

From the viewpoint of soil formation, the important classes of vegetation are forest trees, grasses, and desert shrubs. Trees occupy the humid areas, short shrubs are in the arid lands, and native grass occupies the intermediate zone.

Accumulation and distribution of organic matter and nitrogen in the soil profile are greatly affected by the type of vegetation. Grass is noted the world over for its ability to build a high organic-matter supply throughout the entire surface soil. Under virgin forest the leaf fall tends to raise the organic content of the top horizon. Desert shrubs are not effective in building or maintaining the organic-matter supply in soil.

Jenny [2] has pointed out that the research data at hand indicate that a deciduous forest cover stimulates leaching and accelerates soil-profile development. The high retentive qualities of residues from grass vegetation result in considerably less leaching and a slower rate of soil formation.

The contributions of the larger animals to soil formation are essentially mechanical in nature. Burrowing animals effectively mix and distribute the material of the horizons.

Worms, particularly earthworms, are very active in many soils and greatly affect soil formation as they burrow through the soil and subject the organic and mineral matter to digestive processes. Earthworms drag leaves and blades of grass down into their burrows. Their channels serve as routes for water and air into the soil. The tremendous chemical action of earthworms on soil is evidenced by the fact that soil after passing through a worm's body has a higher available content of several important nutrient elements than before.

**Fig. 2-1.** Even bluegrass provides no guarantee against poor soil structure. Heavy motorized equipment, used without regard to soil moisture conditions in this Indiana apple orchard, has compacted the soil between the rows of trees. The bluegrass roots are very shallow because the soil is tight and platy. (*Kohnke and Bertrand*)

In certain areas of the world much mixing of soil through the profile and exposure of subsurface material to the environment is brought about by ants and termites. Ant activity is greatest in arid and semiarid areas.

Another animal of importance in certain soils is the crayfish. This animal bores numerous vertical channels which are effective in drainage and aeration of some swamp soils.

Man's influence upon soil-profile development can be very pronounced. Fertilization, use of grasses and deep-rooted legumes, and tillage are merely a few of the means at our disposal for deepening the profile. The decomposition of large quantities of organic matter results in the formation of larger amounts of carbon dioxide and organic acids and brings about more intense weathering of the mineral particles of the soil.

The various forms of tillage, while needed for the preparation of land for planting crops, for the control of weeds, and for the regulation of air and water in the soil, have also several distinct dis-

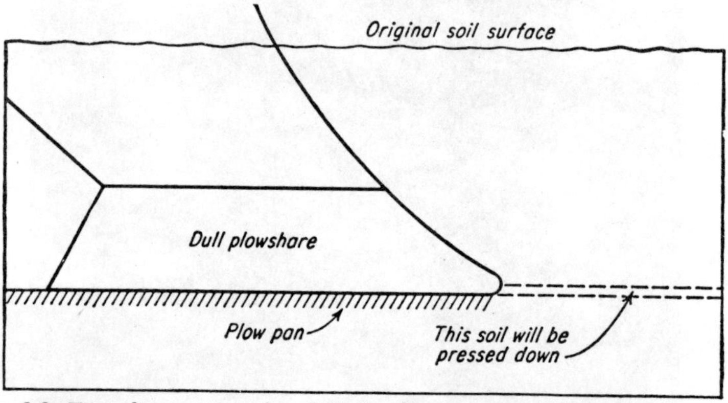


**Fig. 2-2.** Troweling action of a dull plowshare. Since the point of the share is above its base, the soil will be cut above the base and a layer of soil—indicated by the two dashed lines in front of the share—will be troweled down to form a plow pan. When the plowshare is sharp, this will happen to a much smaller extent.

advantages. The loosening of the soil encourages increased aeration and therefore causes the rapid oxidation of organic substances. The pressure of all farm equipment compacts the soil, squeezing the particles together and reducing the pore space, especially the larger pores. This is particularly serious when the soil is wet (wetter than pF 2.7 for mineral soils of medium texture). Plowing can also be responsible for a marked deterioration of soils. If the soil is wet and the plowshare not sharp, the soil will be troweled down into a plow sole.

Erosion exposes lower horizons and brings about more rapid weathering. In the last analysis every farming operation in the field has some effect on the soil profile, and it is up to the alert farmer to direct these changes to his own benefit.

## *Topography*

Topography is considered as a factor in soil formation because of its influence on soil temperature, runoff, and vegetative cover. Vegetative cover has been discussed under the biotic considerations, and temperature is included in the discussion of climatic effects. Therefore the present discussion will deal principally with the influence of relief on the water regime of the soil, which is of particular interest to the conservationist.

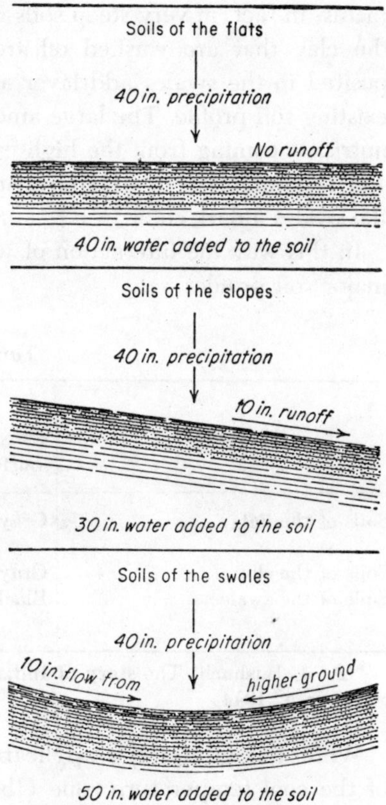


**Fig. 2-3.** Schematic representation of the effect of topography on soil water.

The leaching out of materials from the soil depends on the amount of water percolating through the soil and on the chemical condition of the water. The topography determines to a large extent the amount of water that is available for soil development and plant growth. On a nearly level soil of adequate infiltration capacity all the precipitation water that is not immediately evaporated back into the atmosphere becomes soil water. On sloping ground some water runs off, and this water accumulates in the "low spots."

These illustrations show that more water is available for soil development in the swales and less on the slopes than on level land. The decomposition of the rocks that form the parent material of our soils results in formation of sand, silt, and clay. Where the water penetrates the soil vertically, it carries the clay and the clay-forming substances downward and deposits a clay layer at some depth below the surface. This layer is variously known as a "clay pan," a "hard pan," or the *B* horizon. Eventually this layer may become so dense that it can hardly be penetrated by water. Where water on sloping land moves sideways above ground (surface runoff) or below the ground (lateral seepage), some of the clay formed in the soil is carried off to lower-lying areas. Therefore much less clay accumulates in these soils, and their texture is generally coarser. The steeper the slope, the less clay remains; in fact, in very steep soils no clay layer exists. The minerals and the clay that are washed off from the surrounding slopes and deposited in the swales add layer after layer of fine soil material to the existing soil profile. The large amount of water together with the plant nutrients coming from the high ground results in abundant vegetation and accumulation of organic matter. This "humus" gives the soils of the swales a dark color.

In this way the interaction of water and topography results in three major soil profiles.

*Table 2-1*

| Name | Color | | Indiana Soil Survey* |
|---|---|---|---|
| | Surface soil | Subsoil | |
| Soils of the flats | Gray | Mottled gray and rust brown | II |
| Soils of the slopes | Gray brown | Reddish brown | IV |
| Soils of the swales | Black | Mottled gray and rust brown | VIII |

* T. M. Bushnell, The story of Indiana soils, *Purdue Univ. Agr. Expt. Sta. Special Circ.* 1, 1944.

While this grouping is applicable particularly to the humid sections of the cool temperature zone (the gray-brown podzolic soils), it also represents quite well the soils in other areas where water and topography create differences in soil profiles. There are, of course, soils that do not fit into this simple scheme. One group is that which is developed on material that is so permeable for water that no clay accumulation occurs. Extremes in climates will also prevent formation of soils of these types. By and large, however, this classification is useful in recognizing the three main types of soil profiles that result from differences in topography and water.

## Time

The rate of soil formation varies greatly according to the different factors that contribute to it. It has taken nature many thousands of years to develop most of the soils we have today. Soil profiles 3 feet deep have formed in the temperate zone from the broken-up rock material that the glacier left an estimated sixteen thousand years ago.

*Table 2-2. Rates of Soil Formation from Calcareous Glacial Till in Indiana*

| Soil-material source | Soil series | Age, years | Depth of profile, in. | Average soil formation, years per in. | Average rate of soil formation, lb per acre per year |
|---|---|---|---|---|---|
| Illinoian glaciation........ | Clermont | 300,000 | 140 | 2143 | 140 |
| Early Wisconsin glaciation. | Delmar | 20,000* | 55 | 364 | 825 |
| Late Wisconsin glaciation.. | Bethel | 16,500* | 33 | 500 | 600 |
| Latest Wisconsin glaciation | Nappannee | 13,000* | 30 | 433 | 690 |
| Fill for "Interurban" near Lafayette, Ind.......... | ......... | 80 | 1.5 | 53 | 5600 |
| Spoil bank of coal strip mine, Centenary, Ind.... | ......... | 25 | 1.0 | 25 | 12,000 |

* Age estimates of these soil profiles based on R. F. Flint and Meyer Rubin, Radiocarbon dates of pre-Mankato events in eastern and north central America, *Science*, **121**:649–658, 1955.

Soils derived from solid rock develop much more slowly, while on some of the ash falls of the volcanoes, soils of considerable thickness develop in a few centuries. Flood material may be soil the very day the water has drained off. Soil is not formed inch by inch, but a simultaneous change occurs throughout a considerable depth of the profile, dissolving and leaching chemicals from the top horizons and depositing these in deeper horizons or removing them from the profile. The rate of soil formation is much faster near the surface than farther down where the effects of the weather are not so pronounced. The thicker a soil becomes, the slower the increase in thickness. A definite accumulation of organic matter 1 in. thick has been observed on glacial debris that had been exposed only twenty-five years ago. Soil profiles 2 in. thick have been found on glacial material deposited by man one hundred years ago.

On porous material, soil profiles form faster than on tight material, because more water percolates through them. On a sand dune, stabilized less than one hundred years ago, a podzol profile of 7 to 10 in. in depth was found in a pine forest.

From information concerning the age and depth of soil profiles we can calculate the average rate of soil formation [3]. Such data are summarized in Table 2-2. They indicate clearly that soil formation slows down as the profile gets deeper, because the influences of the weather are getting smaller as the protective mantle of soil increases.

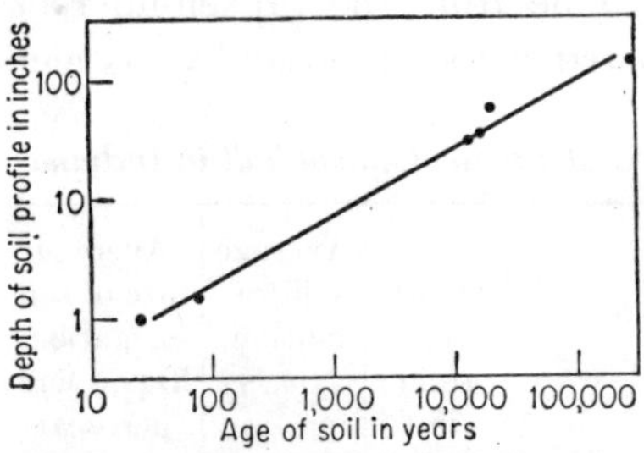


**Fig. 2-4.** Relationship between depth of profile and age of soils, derived from calcareous till in Indiana.

Figure 2-4 shows the total amount of soil formation, as expressed by the depth of the soil profile, to be a function of the age of the soil. This relationship can be expressed by the equation

$$y = At^B$$

where $y =$ the depth of the soil profile, in.

$t =$ the time of soil formation, years

$A =$ the theoretical soil depth, in., after one year (by extrapolation)

$B =$ the ratio between the rate of soil formation at any given time and the average rate of soil formation from time zero to that same time

$A$ and $B$ are constants. In the case used as illustration the numerical values are

$$A = 0.14 \qquad B = 0.561$$

Under conditions of different climate or different parent material this relationship would be shifted, and therefore the magnitude of the constants in the equation would change.

Soil formation on sloping ground is faster than on flat land, while the thickness of soil formed is greater on the flat land than on the sloping land. The explanation for this apparent paradox is that a rather large amount of the soil formed on the slopes is removed by erosion, and the parent material is continuously kept close to the surface, thus becoming exposed to the influence of weathering. On flat ground the soil that is formed remains in place and forms a protective mantle over the underlying layers.

To the soil conservationist this effect of slope on the thickness of the soil is of particular interest. Norton and Smith [4] give this relationship for loessial prairie soils of Illinois. This example shows clearly

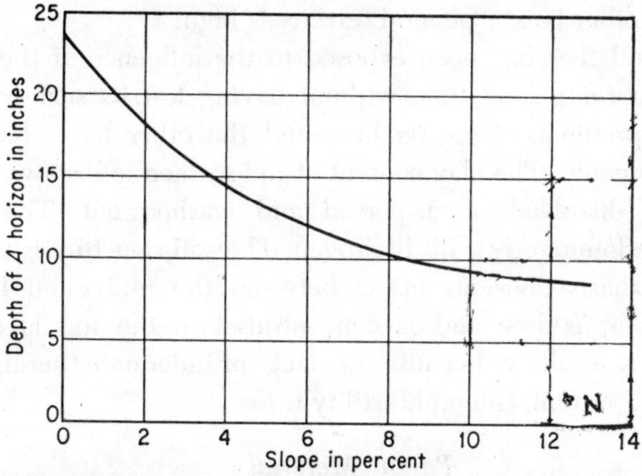


**Fig. 2-5.** Relationship between slopes and thickness of *A* horizons in timbered Illinois soils, derived from loess. (*After Norton and Smith* [4])

how the depth of the topsoil (*A* horizon) decreases as the slope becomes steeper.

Under most conditions in the North Central United States, *A* horizons of virgin soils are no thicker than 10 in. Up to slopes of 10 per cent this thickness is not noticeably affected by the steepness of the slope, but it decreases rapidly as the slopes become steeper.

To evaluate land use and conservation needs properly a conservationist should be able to recognize the stages in soil development.

*Youthful* is a soil that has not been long enough in place to develop a (weathering) profile, e.g., recent alluvial, glacial, volcanic, or aeolian deposits. The fertility status depends largely on the nature of the parent material.

*Immature* is a soil that has had time enough to develop a profile but that has continuously lost enough of the surface so that always new, unweathered portions of the soil have become exposed. It is also a soil in which the weathering processes are slowed down by poor drainage, lack of water, or low temperatures that only an imperfect profile develops.

*Mature* is a soil in which the climate of the region has had a chance to express itself but which has not yet suffered from excessive weathering. A mature soil has a deeper and more distinct profile than an immature soil. It contains a relatively large amount of humus. Lime is removed to shallow depths; clay is predominantly saturated with calcium and other bases. General fertility is high.

*Old* is a soil that has been exposed to the influence of the climate of the region for a long time without having lost its surface soil by erosion. Its profile is deep; its lime and the other bases have been leached out deeply. The clay content of the surface soil is low, because it has been dissolved or dispersed and washed out. The clay is saturated predominantly with hydrogen. The soil structure is generally dense, and sharp divisions occur between the individual horizons. Humus content is low and is concentrated in the top layer. Root penetration is shallow because of lack of adequate aeration and fertility in the subsoil. General fertility is low.

### *Parent Material*

To the conservationist this aspect of pedology takes on a very practical interest because of the influence of parent material on the depth, texture, permeability, and fertility of the profile which is available for use and must be conserved.

Both physical and chemical forces are necessary for production of material in subdivisions small enough to enable soil-forming processes to become active.

All rocks and minerals affect the rate and extent of soil formation, because some decompose readily and others are very resistant. Such minerals as magnetite, pyrite, olivine, and pyroxene may be readily decomposed by oxidation, reduction, carbonation, etc. The end product may be the element in an oxide form, or it may be a clay mineral.

Rocks may be divided on the basis of chemical properties into the following categories: acidic, neutral, and basic.

Upon weathering, acidic rocks, which are high in $SiO_2$, give rise to relatively large particles and therefore have a high proportion of sand. The less resistant portion of the rocks produces fine particles.

These soils are usually light in color because of the low content of iron in the parent rocks.

Neutral rocks are similar to acidic rocks except that they contain a higher proportion of basic materials. They contain more feldspar, which weathers to clay. Therefore soils from neutral rocks contain less sand than soils from acid rocks.

Basic rocks are rich in iron or magnesium silicates, and because of their inherent lack of quartz minerals, they weather into fine particles. These soils are usually red because of the high iron content of the original rock.

## VALUE OF SOIL HORIZONS

The student of soil conservation should have a clear picture of the relative value of the individual soil horizons before he sets out on his task of conserving soil. Practically no soil is uniform throughout a depth of several feet. Exceptions are a few of the water-laid or wind-laid soils. Most soils have a pronounced profile; this means that their properties change with depth.

Typically, the difference between the *A* horizon (the zone of leaching) and the *B* horizon (the zone of accumulation) is that:

The *A* horizon is higher in humus, nitrogen, phosphorus.

The *A* horizon is coarser textured and has a higher percolation capacity.

The *B*, especially the $B_2$, has a higher reserve of exchangeable bases. It is high in clay and has a low percolation capacity.

All the individual soil factors that vary with depth may be combined into two main items, productive capacity and erodibility.

Productive capacity is high and erodibility is low in the surface horizon of all virgin soils. This is because of the accumulation of organic matter and bases. The next horizon is frequently quite erodible and considerably lower in productive capacity, especially in timber soils, but its permeability for water is still high. Where it is not sufficiently protected from water impact and runoff, erosion is serious. But the rather high porosity permits an easy establishment of plants. The *B* horizon contains much clay and a fair amount of available potash, but phosphate and nitrogen contents are low, and organic matter is practically absent. Productivity is therefore low. Permeability for water and roots is low, causing much runoff and poor establishment of plants when the *B* horizon is exposed. Since the aggregates of the *B* horizon are fairly stable, detachability and transportability are low, and erosion is not excessive in spite of ample surface runoff.

The materials of the *C* horizons of soils vary so much that no generalization is possible. Frequently, the productive capacity and the

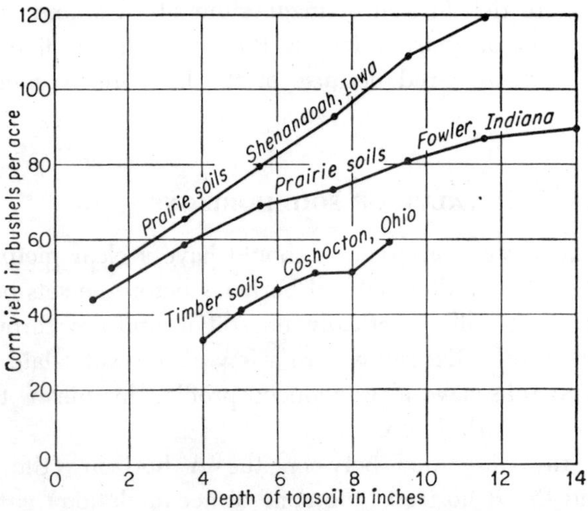


**Fig. 2-6.** The effect of depth of topsoil on corn yield. These data were obtained under conditions of low fertilization. (*After Uhland* [5])

permeability are intermediate between *A* and *B* horizons, but the erodibility may be any value. Since the *C* horizon has lost only little of its components by leaching, it is able to supply bases and minor elements to the plant roots.

These considerations show that in most cases it is wise to attempt to preserve the *A* horizon (ordinary farming practices destroy the $A_0$ horizon). But it is rather likely that the natural combination of the *A*, *B*, and *C* horizons result in better productive capacity and erosion-control value than if the soil were composed of only *A* horizon material. The reason for this is that the slower permeability of the *B* horizon permits the rain water to remain longer in the soil and to be available to plants in larger quantities. Plant roots find some nutrients in greater amounts in the *A* horizon than in the *B* horizon, and vice versa. In an acid soil derived from calcareous material usually only little calcium and other bases are available, but the underlying *C* horizon is rich in these.

The relative importance of the *A* horizon varies with the nature of the soil profile. Where it is underlain by rock or stiff clay at shallow depth, its relative value is great. In the mountainous part of the Scottish Highlands, for instance, forest removal during World War I resulted in denudation of the rocks and in complete destruction of the value of the land. On deep loessial soils, on the other hand, the subsoil is very similar to the topsoil, except for organic matter, nitrogen, and phosphorus, and the relative value of the surface soil is small.

There are also soil types whose *B* horizons furnish a better habitat for plants than their *A* horizons. Some of the old soils of the Piedmont in the Southeastern United States have an *A* horizon consisting largely of silica sand, while the *B* horizon is a mellow loam of mixed minerals. In such a case moderate erosion increases the land value.

Experiments in the Middle West have shown that corn yields increase with the increase in depth of the topsoil (*A* horizon) up to about 12 inches [5].

## REFERENCES

1. Bushnell, T. M.: The story of Indiana soils, *Purdue Univ. Agr. Expt. Sta. Special Circ.* 1, 1944.
2. Jenny, Hans: "Factors of Soil Formation," McGraw-Hill Book Company, Inc., New York, 1941.
3. Kohnke, H.: Interaction between age and depth of soil profiles, *Soil Sci. Soc. Am. Proc.*, in press.
4. Norton, A. E., and R. S. Smith: The influence of topography on soil profile character, *J. Am. Soc. Agron.* 22:251–262, 1930.
5. Uhland, R. E.: Crop yields lowered by erosion, *USDA SCS-TP*-75, 1949.

# 3

# *Soil Erosion*

## GEOLOGIC EROSION

No sooner has a continent been lifted from the sea than water, ice, and wind begin to carry off its surface material. This smoothing down of the hills and mountains, counteracting the great upheavals of the earth's crust, is called geologic erosion. It has gone on since time immemorial and is largely responsible for the present shape of the earth's surface.

### *Forms of Geologic Erosion*

Geologic erosion occurs as leaching, surface erosion, landslides, and oxidation.

**Leaching.** Solution of minerals and organic matter followed by percolation or lateral movement of the dissolved substances is probably the most effective form of geologic erosion. All components of the earth's surface are soluble and find their way eventually into the sea. The solubility of the various rocks and minerals is very different. Gold and silica are very resistant to solution in temperate climate. Calcium carbonate is readily soluble in carbonated water. Therefore some of the most spectacular examples of geologic erosion by solution are the limestone caves.

The concentration of the various ions in river water [14] bears proof of the great amounts of material carried away from the land in solution. Calcium, bicarbonate, and sulfate are removed in the greatest quantities; magnesium, sodium, potassium, nitrate, and chloride represent the bulk of the other ions carried by the rivers. Actually, all compounds in the earth's crust are subject to disintegration and solution.

Carbon dioxide from the air and from respiration and decomposition of plants and animals and organic acids helps in the dissolving action of the percolating waters.

**Surface Erosion by Water.** Surface erosion is the removal of exposed soil or rock. Any part of the earth's surface that is not protected by vegetation is liable to attack by water, ice, and wind. Lack

**Fig. 3-1.** Badlands of South Dakota. Sedimentary rocks in an area where droughts and heavy rainstorms alternate have not produced enough vegetation to protect them from geologic erosion. (*USDA*)

of vegetation may be due to adverse climatic, topographic, or chemical conditions, such as extreme cold, drought, steepness, or alkali or salt concentrations. Other causes may be fire, uprooting of trees, burrowing by animals, and water action along streams, lakes, or oceans. Moving ice, either in the form of glaciers or as ice floes in rivers, can carry tremendous quantities of rock and soil.

For surface erosion to occur, it is not necessary to have the soil completely free of vegetation. The smallest area exposed to the beating of raindrops and to flowing water permits the removal of soil.

**Landslides and Soil Creep.** Wherever soil on a pronounced slope becomes saturated with water, the possibility exists that it will become

**Fig. 3-2.** Erosion cycle. A buried juniper is uncovered by an arroyo. This is typical of many buried forests in Arizona. Present gully erosion is exposing these trees which once grew on a lower level in the valleys. They are important landmarks of past cycles of erosion. (*USDA*)

so well lubricated that gravity will cause it to slide down the hill. This is particularly likely to happen when a fairly pervious mass of soil and subsoil overlies impervious strata such as clay or rock. The water will then accumulate above these layers and float off the material above. Conditions for landslides are especially favorable along streams that cut away the lateral support of large earth masses.

In contrast to the downhill movement of large, deep masses of earth that are called landslides, soil creep is the bodily sliding down of relatively shallow layers of soil. This usually involves areas of from ¼ to 2 acres. The subsoil in the upper part is exposed, while the surface soil piles up at the bottom of the soil-creep area. Sometimes there are several lines of accumulated surface soil, giving the area a wavy appearance. When the slopes are very steep, water is not always necessary to bring about slides. Rock slides and earth slides occur when the contact with the underlying strata has been weakened through weathering.

**Wind Erosion.** Even without the activities of man, wind has contributed greatly to the sculpturing of the earth's surface. Since wind erosion can become an important item only where the ground is not protected by vegetation, most geologic wind erosion occurs in deserts and other arid regions, along the seashore, and in the flood plains of rivers that carry large amounts of soil materials. Examples are the sand dunes that have been blown up along the sea and large lakes and the wind erosion of glacial sluiceways that has resulted in a thick loess mantle over much of the surrounding area.

### *Factors Affecting Geologic Erosion*

Many of the factors affecting geologic erosion are the same as those that affect man-induced erosion.

The most important ones of these are rainfall, wind, steepness of the slope, aspect, erodibility of the surface material, and the amount of the surface area exposed. Others that are of no or minor influence in the case of man-induced erosion are the physical nature of the rock, earth uplifts and earthquakes, the time since the last uplift, and moving ice.

In areas where limestone and sandstone alternate, it can frequently be observed that the sandstone has weathered away much more than the limestone. This results in limestone ledges above the more easily

**Fig. 3-3.** Young mountains. Erosion has not had time to smooth out the slopes of the Tetons. (*USDA*)

**Fig. 3-4.** Old Mountains. These mountains in Pennsylvania are old and have smooth contours with heavily forested slopes, which no longer erode rapidly. (*USDA*)

eroded sandstone. A prime example of intensive geologic erosion is the Grand Canyon of the Colorado River. Brittle sandstone and easily erodible clay shales form a large part of the sides of the canyon.

The uplifts of land masses, as well as earthquakes, are responsible for changed surface configurations with steep slopes and for the exposure of large areas without vegetation.

The longer the time that has passed since these uplifts occurred, the more erosion has smoothed out the surface. The Rocky Mountains are fairly young. Therefore many of their contours are steep and ragged. The Appalachians are older and consequently smoother. Older yet are the Ozarks. Much of the land that once was very mountainous is now almost flat.

Moving ice has an immense power in eroding away rocks and earth. Land that has been worked over by glaciers is usually smoother than unglaciated land. This is due partly to the eroding of the high spots and partly to the filling in of the low spots with the sediment that had been carried by the glaciers. In mountainous territory, glaciers may concentrate in the valleys and gouge these out deeply.

## *Geologic Erosion and Soil Formation*

The speed with which geologic erosion proceeds varies from case to case. We can hardly speak of "norms of geologic erosion" except for specific conditions. Probably the highest rates occur in desert areas where there is little or no vegetation to impede the power with which wind and sometimes water carry away the surface layers. Of particular interest is the effect of slopes of various steepness on the rate of geologic erosion, because this determines largely what kind of soil has developed.

On very steep slopes practically all soil is washed off as soon as it begins to form, and bare rock is at the surface. Where slopes are somewhat less steep, the rates of rock disintegration and soil formation are slightly greater than the rate of erosion. Here shallow *AC*-profile soils exist. This means that a horizon of eluviation (*A*) and a horizon of slightly modified parent rock (*C*) are formed, but that the iron, silica, alumina, and clay that are deposited as the horizon of illuviation (*B*) in fully developed soils are washed out laterally and have no chance to give rise to such a *B* horizon.

Geologic erosion on soils of more gentle slopes (less than about

20 per cent gradient) is slower than soil formation, the rate depending on the rock material. The older these soils, the deeper are their profiles. Mature *ABC* profiles are formed. On very old soils on gentle slopes the original horizon of illuviation (*B*) has occasionally been leached out and another *B* horizon may have formed farther down. On practically level land the rate of geologic erosion is nearly zero. The zone of eluviation is deep, and much of the plant-nutrient supply that originates in the parent material, as well as much of the clay, has been washed out of the root zone of the plants. Therefore such soils are generally low in fertility and poor in structure.

Climatic variations and the nature of the parent material will change somewhat the picture of the relations of the slope to geologic erosion. One conclusion, however, has rather general application: both excessive rates of geologic erosion as well as no geologic erosion at all are undesirable from the viewpoint of soil fertility. Moderate geologic erosion gradually removes the leached surface layers, but it permits the formation of a deep soil profile whose nutrient supply is constantly replenished from the unleached parent material beneath.

## FUNDAMENTALS OF SOIL EROSION BY WATER

### *Essential Components of Erosion*

Soil erosion is the removal of soil from its original location. In most cases the soil particles at the surface adhere to the body of the soil, and therefore they must first be separated from it before they can be removed. Consequently, a more specific definition would be, "Soil erosion is a process of detachment and transportation of soil materials by erosive agents" [18].* Except for very sandy soils these two phenomena, detachment and transportation, are essential components of soil erosion.

Detachment has to precede transportation. But transportation does not always follow detachment. This means that detachment is essentially an independent variable but that transportation depends on detachment. For example, a severe rainstorm hits a cultivated field.

* It may be well to clarify the terms used in this book in connection with erosion: "erosive" means tending to cause erosion (for example, a highly erosive rainstorm); "erodible" means susceptible to erosion (for example, an erodible soil). "Erosiveness" and "erodibility" are the corresponding nouns.

The surface soil is detached by the impact of the raindrops, but before runoff can occur, the rain stops, and because of insufficient water on the surface, the detached soil particles are not transported away from their original location. Erosion—in the sense of our definition—has not occurred.

It may be argued that any detachment must involve a separation, and consequently transportation, of the soil material from the body of the soil. In the example just given, the raindrops have splashed the soil particles back and forth; consequently, they have been "transported." The net movement, however, was zero, since the average location of a large number of particles has not been changed. This consideration shows that it is difficult to define the process of soil erosion rigidly, although the nature of this phenomenon seems to be perfectly clear.

Detachment and transportation are expressed in different units. This is perhaps the best way to draw a distinct line between the two. The amount of detachment is expressed as weight or volume of soil per unit area, for instance, tons per acre. Its rate is measured as weight or volume per unit area per unit of time. In a corresponding way the amount of transportation is expressed as weight or volume moved through distance per area, for instance, ton-miles per acre. Its rate is measured as weight or volume moved through distance per unit area per unit time.

While transportation of soil is a process that can be easily observed, defined, and measured, detachment has no such clear boundaries. In fact the mode of transportation determines whether the breaking loose of a certain piece of soil is the detachment that is the essential first step of the erosion process. If the clod that breaks off is large, a thin sheet of runoff water cannot carry it away, while the same clod in a gully or a stream would readily be transported.

### *Forms of Soil Erosion by Water*

**Sheet Erosion.** Soil erosion by water can take on a variety of forms. The most widespread and probably the most damaging form is sheet erosion. This is the essentially uniform removal of a thin layer, or "sheet," of soil from a given area of land. Normally, the soil is detached by the impact of falling raindrops, not by the flowing water of sheet flow. Every runoff removes some soil. If the surface is smoothed out through tillage, it is difficult to recognize that any soil has disappeared. After this process has been repeated many times, much of the original surface soil is gone, and the farmer grows his crops in the subsoil, which usually is not so good a medium for plant growth as was the surface soil.

**Fig. 3-5.** Sheet erosion. Sheet erosion has taken heavy toll from this wheatland in southeastern Washington. (*USDA*)

**Internal Erosion.** A frequent result of the impact of the raindrops on bare soil is the washing of the soil particles into the cracks and pores of the soil. In this way the soil becomes less pervious to air and water. This internal erosion may not cause permanent damage, since the soil is not removed from the field, and after the structure has been improved, the quality of the soil may be the same as it had been originally. In fact, in some cases internal erosion may actually be of advantage if it washes organic matter into the soil cracks, thus helping to stabilize the aggregates. By and large, however, internal erosion impairs infiltration capacity, and consequently, it increases runoff and surface erosion.

**Channel Erosion.** In contrast to sheet erosion, channel erosion occurs where surface water has concentrated, so that a large mass of water supplies the energy both for detaching and transporting the soil. Channel erosion exists as rill erosion, gully erosion, and stream erosion.

Rill erosion is incipient gully erosion. It usually is the result of water washing down between the rows of a cultivated crop that is planted

**Fig. 3-6.** Channel erosion. Large quantities of water flowing in a channel have detached and transported the soil. The lister furrows which run up and down the hill allowed this damaging accumulation of water. (*USDA*)

up and down a hill or in implement marks and other slight irregularities of the soil surface. If rill erosion continues only for a short while, tillage operations may smooth out the surface completely, so that the resulting soil profile is identical to one that is damaged by sheet erosion.

Gully erosion is channel erosion that washes so deep into the subsoil that the ground cannot easily be smoothed out by ordinary tillage tools. As in the case of rill erosion, gully erosion is caused by temporary runoff during and right after rainfall, not by permanent or intermittent streams. In most cases gully erosion follows sheet erosion, but there are many gullies that have dissected level land where virtually no sheet erosion occurred. Such situations are river-terrace soils, outwash plains, and other places where the topography encourages concentration of runoff water, especially where a sharp increase in slope causes overfalls of water.

**Fig. 3-7.** Rill erosion. Up-and-down-hill cultivation of this unprotected strawberry field in Oregon has led to severe rill erosion. (*USDA*)

Stream erosion is the carrying off of the soil material on the sides and in the bed of a permanent or intermittent stream. The outside bank of a bend in a stream is particularly subject to erosion because the energy of moving water is directed toward it.

There are no distinct boundaries between rill erosion and gully erosion and between gully erosion and stream erosion. The thing they have in common is that detachment is caused by the energy of flowing water, not by the impact of the raindrops.

**Fig. 3-8.** Gully erosion. A main gully with its feeder gullies makes farming impossible on a large part of this once-productive Nebraska field. (*USDA*)

**Fig. 3-9.** Gullies destroyed this field. Soil and subsoil are sandy. Once the gully started, the soil material was so detachable that it was washed away by the water flowing through the gully, undercutting the fairly good sod. (*USDA*)

**Mass Movement of Soil.** Sometimes it is the lubricating action of slowly percolating or even of stationary water, not its kinetic energy, that causes erosion. This comes about where a layer of clay becomes saturated with water and allows the mass of soil to slip downward.

**Fig. 3-10.** Extreme gully erosion. An example of what will happen on much of the steeper land when erosion is unchecked. The subsoil is at least as resistant to erosion as the surface soil. Consequently, V-shaped gullies develop. (*USDA*)

For this movement to occur the following three conditions must exist: (1) There must be a slope sufficiently steep for soil masses to slide; (2) there must be a slowly permeable layer in the soil, some depth below the surface; and (3) there must be enough water in the soil mass to fully saturate the layer just above the impermeable layer.

This layer is usually clay or soil with a high clay content, such as a *B* horizon, but it may also be a rock stratum. Clay shales are frequently the cause for water accumulation and landslides.

The reason that clay is normally the material on which the soil slides is that clay—being a hydrophylic colloid—becomes extremely slippery when wet and consequently acts as a lubricant.

Mass movement of soil occurs in various forms. Landslides of any size represent the breaking away of soil masses that have insufficient support on the downhill side (Fig. 3-11). Small landslides are sometimes called "slips."

**Fig. 3-11.** Landslide. The stream has removed the support of a large mass of unconsolidated material. An unusual amount of rain has caused this huge landslide. (*USDA*)

*Surface creep* is the movement of a relatively thin layer of soil, usually less than 3 feet, over a fairly large area. In flowing downhill this sheet may be pushed together into folds which occasionally break up and give the area a somewhat weird appearance.

*Cattle terraces* are a form of mass movement of soil in which the weight of the cattle supplies the energy for detachment and transportation. In common with landslides and soil creep, cattle terraces are formed mostly when the soil is wet. The cattle paths that result in cattle terraces are nearly horizontal or on a slight grade. Steeper cattle paths also cause movement of soil under the hooves of the animals, but the loosened soil soon washes away to form a rill or gully, leaving no evidence of mass movement.

*Subsidence* of soil is the lowering of the elevation of surface soil due to disappearance or shrinkage of the material below. Since no sideward movement is involved, this may not be considered as erosion in the ordinary sense. It is a form of soil mass movement and is usually the result of shrinking of organic soils caused by drainage. This shrinkage is largely the direct result of dehydration, but it may also be caused by the oxidation of the organic matter.

### *Sources and Forms of Erosive Energy*

*Work* is done to erode soils. Work is applied energy. "When a force acts against a resistance to produce motion in a body, the force is said to do work." This energy in the case of soil erosion is brought about by moving water, wind, or solids; by thermal energy; or by gravity alone.

In the last analysis all erosive agents obtain their energy from the sun, from the gravitation and the rotation of the earth, and from geologic upheavals.

Energy is manifested in two forms, potential and kinetic energy. Kinetic energy is energy due to motion. It is proportional to the mass that is moving and to the square of the velocity of the movement. The kinetic energy of a mass $m$ moving with a velocity $v$ is

$$E_k = \tfrac{1}{2}mv^2$$

Energy can be expressed in ergs, foot-pounds, calories, kilowatthours, or horsepower-hours. Foot-tons is a practical unit to express the impact energy of a rainstorm.

The units of energy can be derived and expressed in the *MLT* (mass–length–time) system as follows:

$$\begin{aligned}\text{Velocity} &= \text{distance per unit of time} = LT^{-1}\\ \text{Acceleration} &= \text{velocity per unit of time} = LT^{-2}\\ \text{Force} &= \text{mass times acceleration} = MLT^{-2}\\ \text{Energy} &= \text{force times distance} = ML^2T^{-2}\end{aligned}$$

Any moving mass possesses kinetic energy. In the case of soil erosion the mass may consist of water, air, or soil.

Potential energy is due to the position of one body with respect to another. The potential energy $E_p$ of a mass $m$ raised through a distance $h$ is

$$E_p = mgh$$

where g is the acceleration due to gravity.

Since gravity is essentially a constant, the potential energy of a given soil mass is proportional to its height above the level to which it can descend.

The ultimate amount of potential energy available to assist in erosion is determined by the height of the soil mass above the base level of erosion. This is the lowest elevation to which it can be transported. For most locations this is the sea. For the individual case of erosion the amount of potential energy depends on the elevation difference between the original position and the foot of the slope.

Generally speaking, sheet erosion is caused by the kinetic energy of moving water—both raindrop impact and surface runoff—while mass movement is largely due to the potential energy resulting from the position of the soil above a point to which it can slide or drop. Both kinetic and potential energy contribute to gully erosion.

## *Detaching Agents*

Soil can be detached from its compound with the ground in a variety of ways. Probably the most important soil-detaching agent is the falling raindrop [20]. The larger its size (mass) and the faster its descent, the greater is its energy and the more efficient a detaching agent the raindrop becomes.

Sheet flow of water is hardly ever an important detaching agent, since its velocity is limited; it is seldom more than ¼ mile per hour (about ⅓ foot per second). Channel flow, however, can supply a tremendous detaching energy, since its velocity as well as its mass may become very large. Channel flow is probably responsible for the largest portion of all detachment in gullies and streams. Together with the energy supplied by moving water, such objects carried by the water assist in detaching the soil. Such objects include ice, floating trees and similar debris, and sand and stones carried as bed load or in suspension. Solid materials concentrate the energy of a relatively

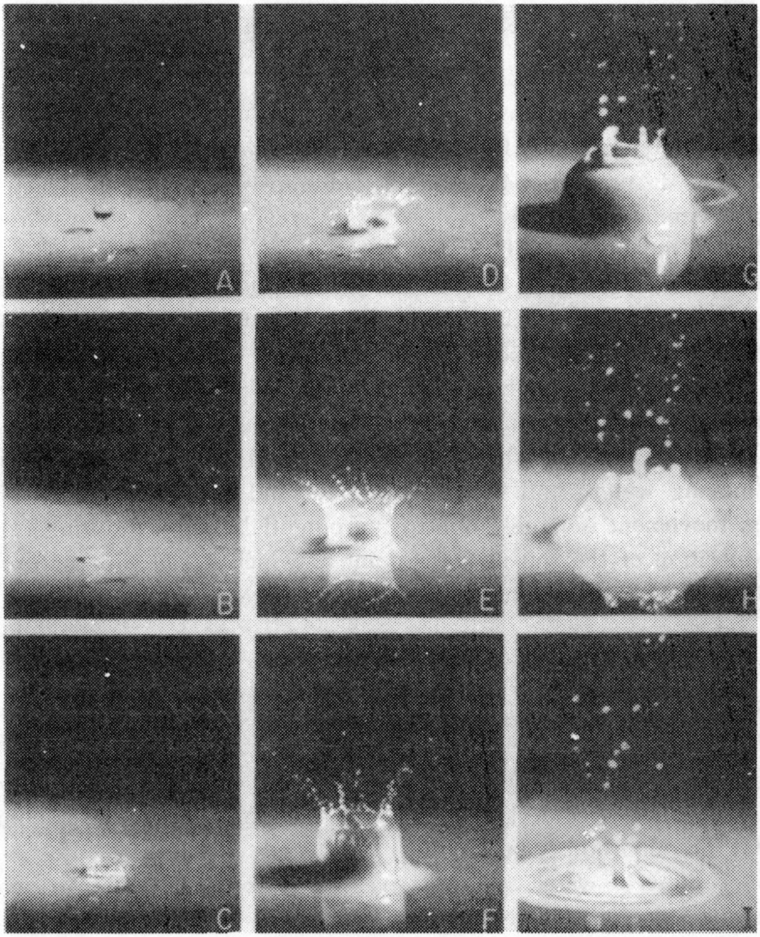


**Fig. 3-12.** Soil detachment by raindrops. The detaching capacity of the rain is shown by the action of a large drop hitting shallow water. Many particles are thrown into the air by this action. (*USDA*)

large mass to a small point and are therefore very effective detaching agents.

The natural process of wetting and drying of soil through the swelling and shrinking of the colloids breaks up soil masses into small aggregates and thus detaches them from the ground. Also freezing and thawing can bring about loosening of soil and may make it prone to erode. In addition, freezing causes a great concentration of water in the surface soil. This reduces infiltration capacity and increases the potential amount of runoff. The greatest erosion hazard exists when the soil at the surface is thawed out but a frozen layer below restricts or even prevents downward movement of water.

Trampling of livestock, especially on hilly land, and the use of tillage implements detach soil. The latter, through excessive aeration of the loosened soil, also causes rapid oxidation of the organic matter, thus destroying some of the bonding material that holds the soil particles together.

### *Transporting Agents*

Flowing water is the most important transporting agent active in soil erosion by water. It transports soil because of its buoyancy and turbulence. Both channel flow and sheet flow transport detached soil particles. Channel flow can easily reach velocities of 4 or more miles per hour and the water is usually in turbulent motion. This means that water that was at the bottom of a stream soon rises to the top and vice versa. In this way any soil that is carried with it is constantly held in suspension and even larger stones are rolled along the ground as bed load. The erosive power of a stream is proportional to the square of its velocity and to the mass of water per unit time. Hence for the same mass of water per unit time (amount of flow), a stream with a small cross section is much more erosive than one with a large cross section. In the case of sheet flow the speed of the water is so small and the depth of it is so minute that no turbulence exists unless it is stirred up by falling raindrops.

Stokes' law expresses the rate of settling of a rigid particle in a liquid.

$$V = \frac{2r^2 (d_s - d_1)g}{9y}$$

where $V$ = velocity of settling, cm per sec
$r$ = radius of the particle, cm
$d_s$ = density of solid particles, per cc
$d_1$ = density of liquid, per cc
$g$ = acceleration due to gravity, cm per sec$^2$
$y$ = viscosity, poises

For soil in water at 20°C this becomes

$$V = 34{,}700r^2$$

This means that the rate of settling of the various soil particles is at about the following rate: fine sand (0.1 mm radius), 4 cm per sec; silt (0.01 mm radius), 2 cm per min; coarse clay (0.001 mm radius), 1¼ cm per hr. Colloidal clay, if dispersed, stays in suspension indefinitely because of Brownian movement.

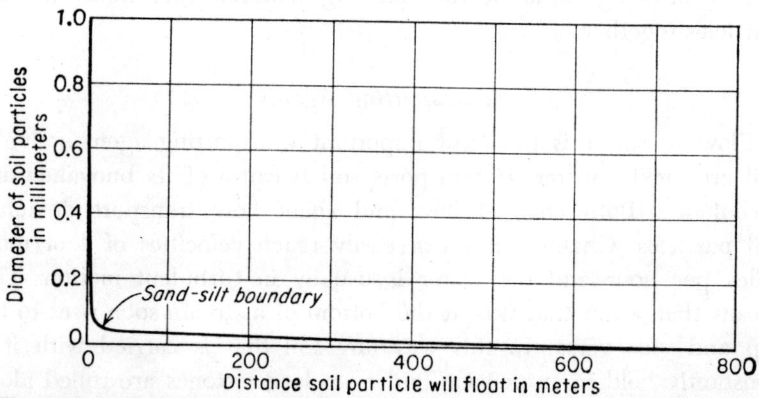


**Fig. 3-13.** Settling distances of soil particles of various sizes. The curve represents the distance a mineral-soil particle will travel in a stream of water of a velocity of 1 mph and a depth of 10 cm before reaching the bottom. This distance is inversely proportional to the square of the diameter of the particle. It is assumed that the particle starts at the surface of the water and that the stream is not turbulent.

Figure 3-13 illustrates that sand in a nonturbulent stream is settled almost immediately. In a stream of 1 mile per hour velocity, fine silt is carried up to 1,290,000 cm (8 miles) before it is lowered 10 cm. Only clay and organic matter would stay in suspension for a greater distance. Consequently, sheet flow would be a very inefficient transporting agent for soil if it were not for the fact that normally, during sheet flow, rain is falling.

The impact of each raindrop stirs up the water and even penetrates a thin sheet of water and loosens—detaches—the soil under the water. In this way the soil is constantly brought to the surface of the sheet and is carried downhill by the moving water. This combination of the detaching effect of raindrops and the transportation of constantly agitated soil particles in sheet flow is responsible for the greatest amount of soil erosion.

Rain splash may in itself be a transporting agent. While the net movement of soil by rain splash that falls vertically on level land is zero, the situation changes if either the raindrops fall sideways, because of wind, or if the land is sloping. Under such conditions splash may cause a considerable amount of erosion.

Transportation of soil is also brought about by livestock, by tillage equipment, or by gravity in the case of mass movement of soil.

### *Estimating the Erosion Hazard*

In order to arrive at an estimate of the rate of erosion that may occur under a given set of conditions, we have to get a picture of the detachment hazard and of the transportation hazard. Since only detached soil can be carried away, erosion can never exceed the rate of detachment. Under some conditions the rate of detachment may be the factor limiting erosion. This would be the case where a very cohesive soil lies on a steep slope and a heavy rain causes sheet flow over the area. Practically all the loosened soil would be washed off the field.

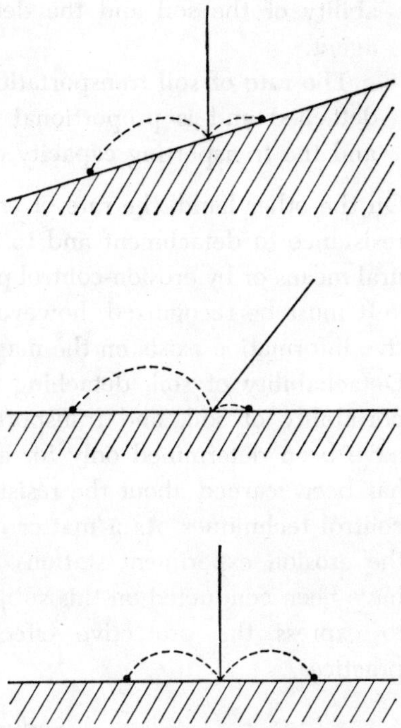

**Fig. 3-14.** Soil transportation by rain. Rain falling vertically on a sloping field or driven by a wind on a level field has the tendency to move the soil downhill or in the direction of the wind, as is indicated in the schematic drawings. Where rain falls vertically on level land, the net movement is zero.

On the other hand, a large amount of a coarse-textured soil with little clay or organic colloid to hold it together may be detached by a heavy rain. But if this soil lies on fairly flat ground, not much of the detached soil is transported away, because the runoff velocity would be slow and the coarse particles would settle out quickly and could not be carried far. Many other combinations of detachment and transportation conditions could be imagined readily. All these situations might be summarized by the following statements:

> The quantity of the soil detached is proportional to the detachability of the soil and the detaching capacity of the detaching agent.
>
> The rate of soil transportation is limited by the amount of soil detached and is proportional to the transportability of the soil and the transporting capacity of the transporting agent.

On the other hand, the rate of erosion is inversely proportional to the resistance to detachment and to transportation that is caused by natural means or by erosion-control practices [18].

It must be recognized, however, that to date, hardly any quantitative information exists on the magnitude of the factors just mentioned. Detachability of soil, detaching capacity of detaching agents, transportability of soil, and transporting capacity of transporting agents have been determined only for a few isolated situations. Much more has been learned about the resistance to erosion provided by erosion-control techniques. As a matter of fact, a large number of studies on the erosion experiment stations in many parts of the United States have been conducted on this subject, and factors have been developed to express the protective effects of the various soil-conservation practices.

## *Sedimentation*

**The Sedimentation Process.** All soil that is eroded from its original location is deposited at some other place. Sedimentation is an integral part of the erosion process. Just as we distinguish between geologic rates of erosion and accelerated erosion, we distinguish between geologic rates of sedimentation and accelerated rates of sedimentation [42].

The distance eroded soil particles travel varies greatly. It may be a few inches, or it may be hundreds of miles. Frequently the coarse-sand fraction of eroded soil is left in the depressions between the rows of cultivated crops while the finer material is kept in suspension until the velocity of the runoff water has slowed down, so that first the fine sand and later the silt is dropped out [45]. Clay and humus stay in suspension as long as they are not coagulated. Dispersed, they are so fine that the molecular motion of the water will keep them in suspension. Coagulation of these soil colloids results usually from the presence of a specific concentration of electrolytes (usually salts) in the water. If the water is entirely still, as it is in a lake, even fine clay will settle out if it is only slightly coagulated. Wherever clay suspensions come in contact with water of high salt concentration, as they do in the ocean or in the lakes in arid or semiarid areas, the clay coagulates quickly and is sedimented out very soon.

**Effects of Sedimentation.** Sedimentation makes erosion an intermittent process. Soil that has been deposited at the bottom of the slope because the velocity of runoff decreased or because the runoff decreased in volume or stopped altogether may be picked up again during the next storm and carried an additional distance.

Each time, the fine materials are carried farthest, and the stones and the sand drop out first. This combination of erosion and sedimentation causes a textural separation of the soil that is eroded from its original position. The rocks are left in the upper, steeper reaches of the watershed and the sand is carried farther but is frequently deposited on or near the creek banks. It is a common observation that the valley soil immediately adjoining a creek is sandy, while farther away from the creek near the valley wall, it is fine-textured. The sandy material near streams is generally deposited in the form of overflow dikes as the flood waters slow down and thus lose their carrying capacity when they leave the main stream and spread over the bottomland in a thin sheet. The finest particles settle out only in standing water. Consequently, the depressions in a flood plain are usually covered with clayey soils, because here the flood waters cannot escape.

**"Silting Up" of Reservoirs.** One of the results of sedimentation is the filling up of reservoirs that are placed in the path of a river or stream. The water in the reservoir is very nearly still, and therefore a part of the soil that the stream has carried is dropped out. While the situation varies from case to case, generally much of the large material, rocks and coarse sand, is either dropped out before the stream enters the lake or just at the entrance of the stream into the lake, building a debris cone. This is particularly true in the smaller headwater streams with steep gradients. A good deal of the clay and organic matter remains in suspension—at least in the smaller, nonsaline reservoirs—and is carried out with the water at the lower end of the lake. This means that the material that is forming the sediment over the largest area of the reservoir consists mostly of silt and fine sand, intermixed with some clay and organic matter. In such sediments, silt frequently makes up more than half the bulk of the material, thus giving rise to the name of "silt" for the entire deposit.

**Fig. 3-15.** A debris cone filling a channel. A debris cone has developed at the confluence of the ditch with the larger creek in the foreground. (*USDA*)

The rate at which reservoirs are filled up with sediment depends on the erodibility of the soil in the contributing watershed, its topography, climate, type of agriculture, and the ratio between the size of the watershed and the volume of the reservoir. In some cases these factors combine to cause a reservoir to be filled up completely within a few years [5].

The rate of erosion in the watershed is no reliable index of the sedimentation of a reservoir. In small-headwater streams the ratio between erosion and sedimentation is particularly high. In a large watershed much of the coarse erosional debris may be settled out in the stream valley long before the water enters the reservoir. The larger the watershed, the smaller is the proportion of the eroding soil that reaches the reservoir. Bank erosion can contribute a sizable portion of the material carried by a stream and in some cases is responsible for the major part of the damage done by sedimentation. According to Eakin [17] silting in of reservoirs in the United States occurs at rates between 0.000082 and 0.056 inch of soil from the entire watershed per year. In isolated cases the rates of filling may be considerably faster, but seldom where the watershed is larger than 100 square miles. While these figures seem small, the rate of filling up of reservoirs may be very high, especially where a watershed in hilly erodible land sends its sediment-laden water into a small reservoir without allowing the water to slow down for a longer distance in a valley. Such reservoirs sometimes fill up completely within fifty years or less.

Even large reservoirs cannot maintain their capacity indefinitely. During the first fourteen years after Hoover Dam was completed, 2 billion tons of sediment was deposited in Lake Mead, reducing its storage capacity by 3.2 per cent [64]. The ratio of the watershed area to the reservoir capacity is an important item influencing the rate of filling of the reservoir. If this ratio, in acre-feet reservoir capacity to square miles of area, is small—less than 50 to 1—the useful life of the reservoir will be short. If the ratio is large—1000 to 1 or more—it will take a long time before the reservoir is filled [28]. Lake Calhoun in northern Illinois, a reservoir constructed in 1924 for recreation purposes, is an example of the rapid loss of water-storage capacity where the drainage area is relatively large. By 1936 the reservoir had lost 35.7 per cent of its original capacity to sediment. By 1947 only 26.4 per cent of the water-storage capacity remained, and the lake was essentially useless [62]. This is in an area of predominantly fertile prairie soils and gentle slopes. The high rate of erosion is probably due to the fact that 80 per cent of the drainage area was in cultivation, mostly in row crops.

**Sedimentation and Floods.** Wherever the rate of erosion is increased over geologic norms, sedimentation is increased also, and the soil is deposited in or near the stream. Under undisturbed natural conditions the silt that deposits in the channel is largely washed out by one of the following floods or as bed load during normal stages, and only a slow adjustment of the course of the river takes place. Under conditions of accelerated sedimentation the river bed builds up quickly with coarse material that cannot be washed away readily, especially if it becomes covered by vegetation. The result is that the cross section of the bed decreases and there is not enough space for the water in case of increased runoff. The water flows over the banks, flooding the lowlands more frequently than when the channel is not restricted by sediments.

**Damages and Benefits from Accelerated Sedimentation.** Some of the damages from sedimentation have been mentioned already. The filling up of reservoirs and the resulting loss of capacity and usefulness are a very serious effect of sedimentation. Reservoirs are always placed in the most suitable and most economical location, and once they are filled up, an irreplaceable asset has been lost forever, since no feasible method exists to remove the sediment from a filled-up reservoir. The water supply, the electric-power potential, and the recreation value of the reservoir are gone. It is seldom possible to find an equivalent location nearby.

Raising the bed of rivers causes an increased number of detrimental floods. Men have built dikes to hold rivers in place, but during severe runoff periods these are unable to confine the water. One of the classic examples of the damage caused by sedimentation of soil in a stream is the Hwang Ho River in China, which has broken its dikes and shifted its course many times during the last four thousand years, causing untold losses in human lives and harm to property [65]. Raising the bed of a river has other serious consequences. The level of the water is raised in the river as well as in the alluvial land of the river valley. Frequently, alluvium has naturally a high ground-water table. If it is raised, the water level may come too close to the surface, so that the land becomes swampy and useless for agriculture.

Coarse sediments in a river cause much of the water to flow below the bed, restricting the amount of surface water. In this way its carrying capacity is reduced, and more sediment will be deposited in this stretch of the river, thus increasing the damage [66].

Sedimentation on agricultural land may be beneficial or detrimental. This will largely depend on the texture and fertility of the sediment and on the rate of its deposition. If the soil material that is deposited is of medium or fine texture and if each time so little is deposited that ordinary tillage can readily mix it with the previously existing surface

**Fig. 3-16.** Detrimental sedimentation. Rich black prairie soil in Illinois has been covered with 24 in. of less fertile soil from eroding slopes.

soils, sedimentation generally improves the land by adding fertile material. On the other hand, covering the alluvium with a thick layer of coarse sand or even stones is obviously damaging. Accelerated sedimentation is generally detrimental, while geologic rates of sedimentation are usually beneficial.

## FACTORS AFFECTING SOIL EROSION BY WATER

In causing soil erosion, water plays a variety of roles. Both in detachment and transportation, water is the main agent. To bring about erosion, water must be in motion.

## *The Hydrologic Cycle*

The continuous movement of water in nature from the sky to the ground and to the sea and back again to the sky is called the *hydrologic cycle* or the *water cycle.* Actually there are several cycles, since water follows various routes as is indicated in Fig. 3-17.

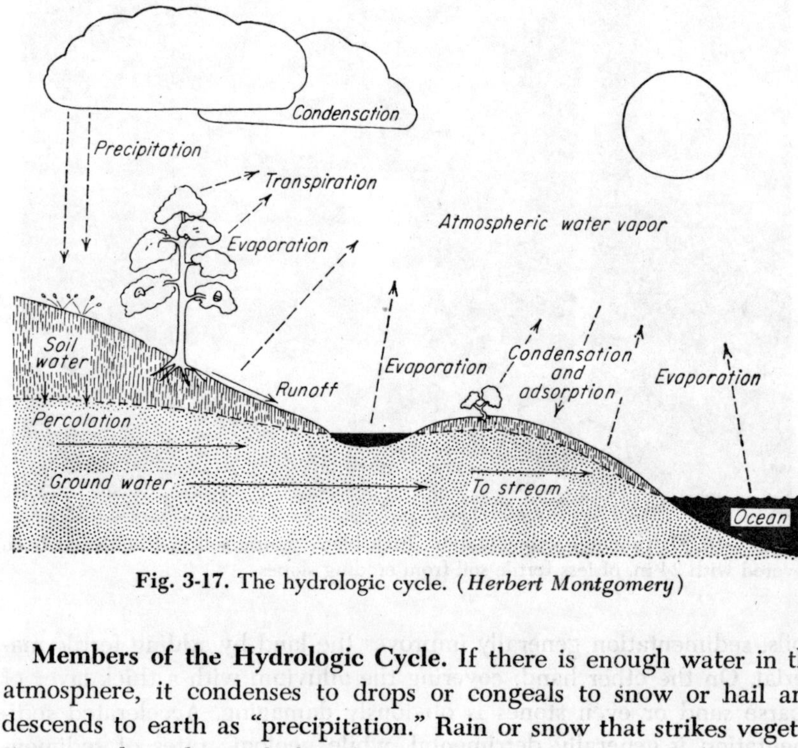


**Fig. 3-17.** The hydrologic cycle. (*Herbert Montgomery*)

**Members of the Hydrologic Cycle.** If there is enough water in the atmosphere, it condenses to drops or congeals to snow or hail and descends to earth as "precipitation." Rain or snow that strikes vegetation or other objects before hitting the ground is said to be "intercepted." The entering of liquid water into the ground is "infiltration," and its continued downward movement is "percolation," which contributes to the "ground water." The water that remains on the surface may be caught in low spots as "surface storage," or it may flow off laterally as "surface runoff," or it may be returned to the atmosphere as "evaporation." Water that enters the plants is given off again as "transpiration." If the conditions are favorable, atmospheric water may be added to the ground through "adsorption" or "condensation."

**The Water Equation.** In the study of soil erosion and soil conservation we are concerned mostly with rainfall and runoff, since they are respectively the most important detaching and transporting agents. The water equation expresses the relations between these two phenomena and the other members of the hydrologic cycle for a given period of time and for a specific mass of soil.

Water gain — water loss = water storage (either positive or negative)

The components of this equation are given below:

| WATER GAIN | WATER LOSS | WATER STORAGE |
|---|---|---|
| Precipitation<br>  Rain<br>  Snow<br>  Hail<br>  Sleet<br>  Fog<br>Condensation<br>  Dew (plants)<br>  Condensation (soil)<br>Adsorption | Runoff<br>Percolation<br>Evaporation<br>Transpiration | Change in soil moisture content<br>Interception storage<br>  Transitory interception<br>  Residual interception<br>Surface storage<br>  Depression storage<br>  Surface detention |

Note that infiltration is not shown in the enumeration of water losses. This is because the equation refers to a specific mass of soil. Infiltration represents the main reason for "change in moisture content of the soil" during rainfall.

This equation can also be written:

Water gain — water storage = water loss

Condensation, adsorption, percolation, evaporation, and transpiration occur seldom at rates higher than a few hundredths of an inch per hour, while rainfall, runoff, and water storage can occur at rates of several inches per hour. Therefore the first five items can be neglected in a calculation to estimate the amount of runoff from a rainfall of short duration. The equation becomes:

Rainfall — water storage = runoff

This shows that surface runoff can only be controlled by high infiltration capacity of the soil, by surface depressions, and by a heavy cover of plants or plant residues. By far the most important of these three is high infiltration capacity of the soil.

## *Precipitation*

The word *precipitation* means the dropping down of some object. In hydrology it means the dropping of water in the liquid or solid state to the earth.

**Fig. 3-18.** Erosion pedestals. Leaves and flat stones have protected part of the soil from the impact of rain drops and erosion, while the unprotected soil is detached and moved away. Such spectacular pedestals are usually formed in areas sheltered from wind, which would cause undercutting, and also where there is only limited surface runoff. (*USDA*)

**Forms of Precipitation.** Precipitation may occur as rain, snow, fog, hail, or sleet. Fog can only be considered to be precipitation if it strikes either the ground or the vegetation. Otherwise it may drift off or be evaporated before it can add its moisture to the earth.

Fog is similar to rain, except that the water droplets are so small that they do not fall down. If a very slight breeze moves fog past trees or other plants, the water droplets are held by the branches and the leaves, and when enough water is collected, some of it will drip off. This is called "fog drip."

Generally only rain is of major direct importance in causing erosion. Under some conditions snow melt can be responsible for severe erosion damage. If the snow has partially melted, its thermal value is low. Water is held in the pores of the remaining snow. Very little solar energy, warm wind, and rain are required to melt the remaining snow and to release the water that was held in the pores. The other forms of precipitation are only of limited significance in causing erosion.

**Rainfall Characteristics.** In order to get an idea of the erosive properties of rain the following characteristics have to be considered: intensity (rate); duration; total amount of rainfall; size, velocity, and shape of raindrops; kinetic energy of the rain; and seasonal distribution of rain.

Each of these factors may have a very important influence on erosion. In striking the ground the raindrops supply the main energy for soil detachment. Most of the rainfall characteristics mentioned have a bearing on the resultant runoff, which is the outstanding soil-transporting agent. The factor that has probably the most important effect in runoff is the total amount of rainfall during a storm.

Rainfall intensity is expressed as inches per hour or millimeters per minute. It may be classified as follows:

*Table 3-1. Rainfall Intensity*

| Rate, in. per hr | Descriptive term |
|---|---|
| Less than 0.25 | Gentle |
| 0.25–0.50 | Moderate |
| 0.50–2.0 | Heavy |
| Over 2.0 | Severe |

It must be remembered that while rainfall intensity is expressed in inches per hour, the individual occurrence of a given rate may last only a very short time. For instance, during a quick burst of 3 in. per hr for 2 minutes, actually only 0.1 in. of rain falls. This is the reason that intensity alone gives no indication of the amount of runoff or of erosion caused by a storm.

The definition of an "excessive storm" by the U.S. Weather Bureau takes into consideration both intensity and duration of the rain. An excessive storm has a rate of at least

$$\frac{0.20 + 0.01T}{T} \times 60 \text{ in. per hr}$$

where $T$ is the duration of the rain in minutes [74].

According to this definition a rain of 5 min duration and 3 in. per hr intensity would be an excessive storm. Since the total amount of water is only 0.25 in., little if any runoff would result. It is therefore suggested to modify in the following manner the definitions of an excessive storm when considering its effect on runoff:

A storm of a duration of less than an hour is excessive, if its total amount exceeds 0.80 in.; however, a storm of a duration of more than an hour is excessive only if its rate exceeds

$$\frac{0.20 + 0.01T}{T} \times 60 \text{ in. per hr}$$

where $T$ is the duration of the rain in minutes [74].

Table 3-2 and Fig. 3-19 show the relation between duration, intensity, and total amount of precipitation of an excessive storm.

*Table 3-2. Minimum Characteristics of "Excessive Storms"*

| Duration of rainfall, min | Average intensity of rainfall, in. per hr | Total amount of rainfall, in. |
|---|---|---|
| 8 | 6.00 | 0.80 |
| 20 | 2.40 | 0.80 |
| 40 | 1.20 | 0.80 |
| 60 | 0.80 | 0.80 |
| 120 | 0.70 | 1.40 |
| 180 | 0.67 | 2.00 |

It is recognized that a single definition of an excessive storm cannot possibly separate the storms that cause large amounts of runoff from those that result in insignificant or no runoff. Other factors such as the amount of antecedent rainfall, the permeability of the soil, and the ground cover, to mention but a few, also have a large influence. Nevertheless, the use of a definite yardstick can be very helpful in comparing the effects of rainstorms.

The duration of rainfall is extremely varied. It is measured from the first raindrops until a rain-free period of at least 1 hr. For this purpose a period of 1 hr with less than 0.01 in. of total precipitation is considered rain free.

The amount of rainfall is the volume of water falling on a given area. Therefore it might be expressed in acre-inches per acre or after cancellation in inches or also in millimeters. The amount may refer to an individual rainfall or any specified time period such as a day, a month, a season, or a year.

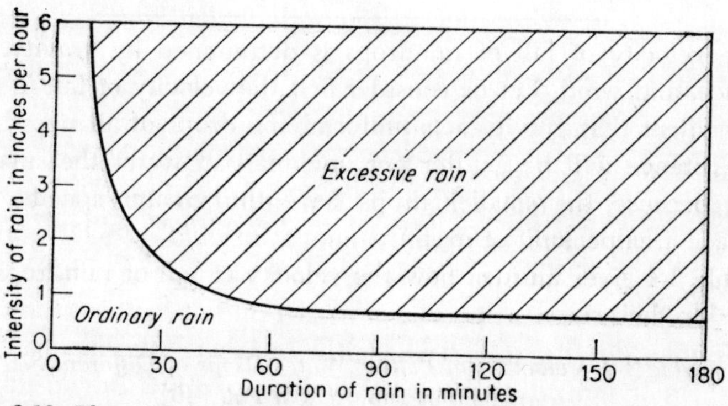


**Fig. 3-19.** The minimum intensity of an excessive rain. A rainstorm is rated as "excessive" on the basis of its intensity and duration. The curve indicates the boundary between an ordinary rainfall and an excessive one.

Raindrops vary in size from very tiny droplets barely larger than fog to a maximum diameter of slightly over a quarter inch (7 millimeters). The majority of the rain water comes down in drops between 1 and 4 mm diameter. In every instant of a rain a large variety of drop sizes occur. Nevertheless, there is a fairly definite correlation between the intensity of rainfall and the median size of the raindrops. This median size—the raindrop diameter dividing the drops of larger and smaller diameter into groups of equal volume—varies from slightly over 1 mm for a rainfall intensity of 0.05 in. per hr to less than 3 mm for a rainfall intensity of 4.0 in. per hr [47]. This shows a threefold increase in drop diameter for an eightyfold increase in rainfall intensity.

The following table shows the relation between rainfall intensity and the median drop diameter [46].

*Table 3-3. Relation between Rainfall Intensity and the Median Drop Diameter*

| Rainfall rate, in. per hr | Median drop diameter, mm |
|---|---|
| 0.01 | 0.75–1.0 |
| 0.05 | 1.0 –1.25 |
| 0.1 | 1.25–1.50 |
| 0.5 | 1.75–2.0 |
| 1.0 | 2.0 –2.25 |
| 2.0 | 2.25–2.5 |
| 4.0 | 2.75–3.0 |
| 6.0 | 3.0 –3.25 |

The velocity of fall of raindrops is determined by gravity, air resistance, and wind. Let us consider first the velocity of fall in still air. It is evident that gravity acts uniformly on drops of all sizes but that air resistance will be greater per unit mass of water the smaller the drop, because the smaller drops have the greater specific surface (surface area per unit of mass).

Table 3-4 gives an idea how the velocity of fall of raindrops is correlated to their sizes:

*Table 3-4. Velocities of Falling Water Drops of Different Sizes after 20 m (65.5 ft) of Fall [46]*

| Drop size mass diameter, mm | Velocity of fall, mph | Velocity of fall, m per sec |
|---|---|---|
| 1.25 | 10.8 | 4.85 |
| 1.50 | 12.3 | 5.51 |
| 2.00 | 14.7 | 6.58 |
| 3.00 | 18.0 | 8.06 |
| 4.00 | 19.8 | 8.86 |
| 5.00 | 20.7 | 9.25 |
| 6.00 | 20.8 | 9.30 |

These data show that the rate of increase in velocity decreases with the increasing size of the drops. Part of the reason for this is the shape of raindrops. Very small raindrops are almost spherical. The large curvature of the surface causes surface tension to maintain the spherical shape of the drop. Larger drops are distinctly flattened out, oblate, with a very flat surface on the lower side. This shape causes a relatively great air resistance. The smaller curvature of the surface of the larger drops causes a relatively small amount of surface tension, and therefore such drops break up as a result of air resistance. This is the reason that water drops of more than 7 mm mass diameter do not exist.

Wind is another factor determining the velocity of descending raindrops. While the maximum velocity of water drops in still air is around 20 mph, winds of much greater velocity exist. Since wind

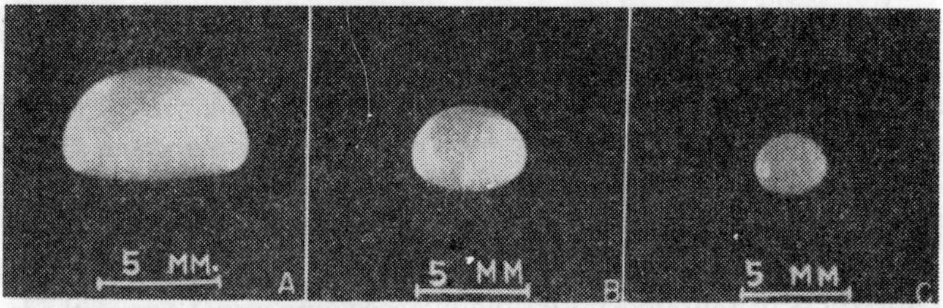


**Fig. 3-20.** Typical shapes of falling water drops. Air resistance flattens large drops. The greater curvature of the smaller drops makes surface tension more effective. They are therefore more nearly spherical.

A (Upper left) diameter = 6.5 mm, velocity = 8.9 m/sec.
B (Lower left) diameter = 4.8 mm, velocity = 8.3 m/sec.
C (Lower right) diameter = 2.8 mm, velocity = 6.8 m/sec.

(*C. Magono, Journal of Meteorology* 11:77, 1954)

movement is horizontal while gravity acts in a vertical direction, the effect of these two factors cannot be added to each other to give the resultant speed. It can be seen, however, that a strong wind can materially contribute to the velocity with which the rain hits the ground.

The most important single rainfall characteristic in the study of soil conservation is its kinetic energy, since this is the prime cause of the disintegration and detachment of the surface soil. It might be assumed that the kinetic energy of rainfall could readily be calculated from the basic formula:

$$E_k = \frac{1}{2}mv^2$$

where $E_k$ is the kinetic energy, $m$ the mass of the moving object, and $v$ its velocity. Since there are drops of many sizes in every rainfall, there are also many velocities that have to be taken into account, a task that is very difficult to perform for a natural rainfall. In order to translate kinetic energy of the falling drops into soil-detaching energy, the size and shape of the drops have to be considered. The size determines the amount of surface tension that tends to retain the shape of a drop when it hits the ground. Small drops are somewhat more rigid than larger ones. The flat shape of the underside of the bigger drops causes a larger proportion of the water to hit the ground all at once.

The seasonal distribution determines to a large extent whether a given annual amount of precipitation will represent a large erosion hazard or not. This depends on the types of precipitation occurring in the various seasons and on the amount of ground cover provided by agricultural crops. Essentially only precipitation that occurs as rain needs to be considered, since snow and fog have no detaching energy and sleet and hail occur only seldom.

Correlations between the season of its occurrence and the potential erosiveness of rain depend on the meteorologic conditions of the area in question. In the American Middle West the monthly amounts of rainfall do not vary a great deal throughout the year [68], but the intensities do, and with them the erosiveness of the rains. Generally the summer rains are of much greater intensities than the winter rains. Erosion studies in Ohio, Missouri, and Wisconsin, for instance, show that more than three-quarters of the soil losses occur during May, June, July, and August [2, 59, 31]. This distribution is caused also by the fact that these are the months when the land is tilled and cultivated and the soil is loose and not protected by crop residues.

Probably the most successful attempt to correlate measurable rainfall characteristics with its erosivity has been made by Wischmeier and Smith [71]. They have calculated the kinetic energy of rain by considering the distribution of drop sizes and drop velocities for each rainstorm intensity.

Figure 3-21 shows the relationship between rainfall intensity and the kinetic energy of 1 in. of rain. Figure 3-22 shows the relationship between rainfall intensity and its kinetic energy per hour. These two graphs are based on the same data. Wischmeier and Smith [71] have found that a good correlation between rainfall characteristics and the resultant erosion from a fallow soil can be obtained when these four variables are added in a regression equation: (1) rainfall energy, (2) the effects of a term measuring the interaction effect of storm energy and maximum prolonged intensity, (3) an antecedent precipitation term, and (4) the cumulative antecedent rainfall energy since last tillage of the soil. By means of such an equation the amount of erosion from a fallow, cultivated soil can be estimated with a high degree

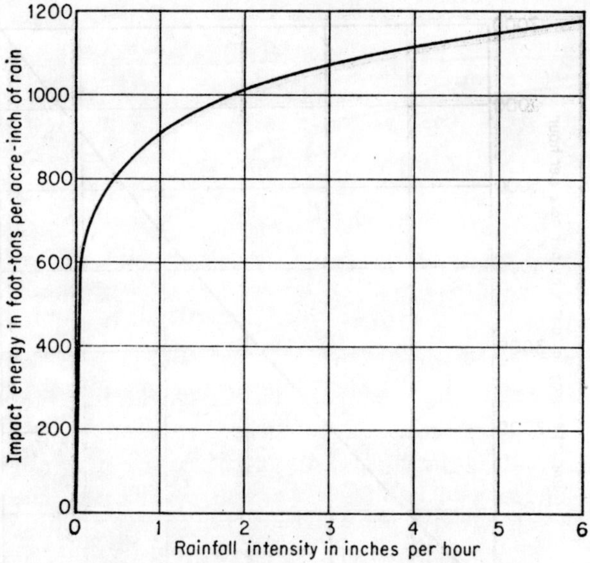


**Fig. 3-21.** Relationship between intensity and impact energy of rain. The predominately small drops of rain of an intensity of 1 in. per hr and less fall fairly slowly and therefore have little impact energy. Above 2 in. per hr, energy impact per acre-inch of rainfall increases only slightly, because there is a limit to drop size. (*After Wischmeier and Smith* [71])

of reliability if the data from a recording rain gage are available and the erodibility of the soil is known.

The kinetic energy of the rainfall in itself affects erosion; however, a closer correlation with erosion is obtained by the rainfall-energy-intensity interaction term. This is the product of the total rainfall energy of the storm and its maximum 30-min intensity. The interaction term "appears to be a good measure of the combined effects of (1) the decreasing infiltration rate during the rain, (2) the geometrically increasing erosion effect of surface flow, and (3) the protection against raindrop splash which is afforded by the film of flowing water." As a matter of fact, this energy-intensity interaction term alone is closely correlated to the actual amount of erosion resulting from a storm. The antecedent moisture index is an expression of soil moisture and consequently is related to the time required for infiltration to reach a minimum rate. The last item in the regression equation, the cumulative antecedent rainfall energy since last tillage, gives a

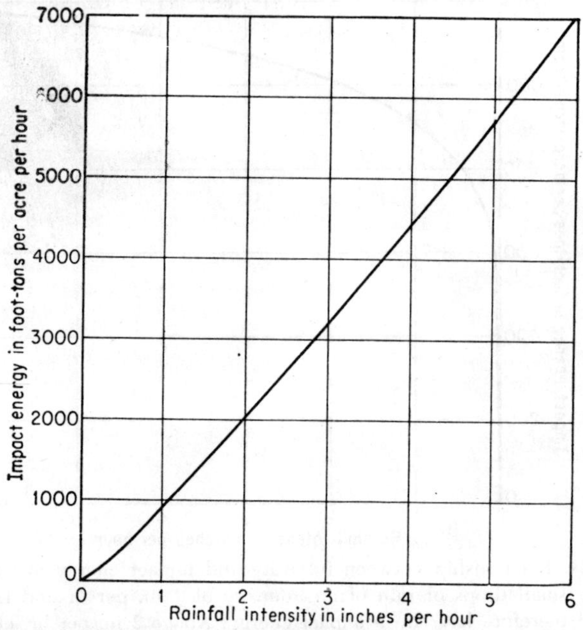


**Fig. 3-22.** Relationship between rainfall intensity and its kinetic energy per hour. In spite of the greater impact energy of rain of high intensity, the relationship between rainfall intensity and rainfall energy *per hour* deviates only slightly from a straight line. This shows the importance of rainfall intensity on its erosion potential. (*After Wischmeier and Smith* [70])

measure of soil compaction, reduction in surface detention due to leveling of the surface by slash erosion and sealing of the soil surface.

In this way the Wischmeier-Smith erosion-regression equation takes into account the direct erosiveness of the rain, the moisture content of the soil at the beginning of the rain, and the physical conditions of the soil surface. And all this is obtained through recording rain-gage data alone. It must not be forgotten that this method of estimating erosion refers to fallow, cultivated land as it exists on the runoff plots of the erosion experiment stations of the U.S. Department of Agriculture.

This excludes vegetation as well as gullies. Plants will obviously greatly modify the erosion resulting from a given storm. Nevertheless the rainfall-energy soil-loss equation is of great value because it permits refinement of the rainfall factor for all locations where recording rain-gage data exist over a period of years. One of its most important

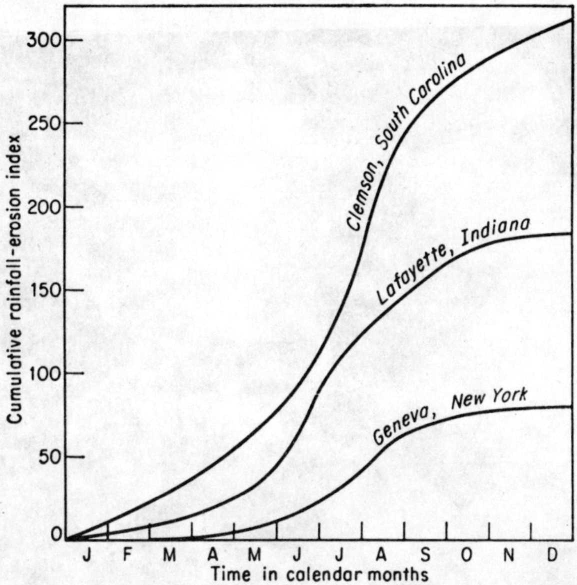


**Fig. 3-23.** Erosion index of rainfall for three stations in the Eastern United States. The cooler the climate, the lower is the erosion potential of rainfall in the Eastern United States. The rainfall erosion index is computed as the product of the rain impact energy per acre times the maximum 30-min intensity for every storm. The summation of these values represents the cumulative-rainfall erosion index. (*After W. H. Wischmeier, A rainfall erosion index for a universal erosion-soil-loss equation* [*unpublished*].)

applications is perhaps the possibility of determining the seasonal distribution of the erosion hazard. Since the energy-intensity interaction term alone correlates well with erosion rates, Wischmeier [72] has used it to calculate the erosion potential of the rainfall throughout the year for a number of locations.

Figure 3-23 shows the accumulated erosion potential for three stations in the Eastern United States. The outstanding difference between the three curves is the total annual amount. The general pattern is similar in that the largest erosion potential occurs in summer. This is probably true in most locations, but by no means everywhere. The Imperial Valley of California is an example of a climate where the greatest erosion potential occurs in winter. As corn is the most widely grown crop that requires cultivation in the Eastern United States, it is well to determine whether the time when the soil is exposed—from

**Fig. 3-24.** Rate and amount of rainfall are recorded on the chart of this rain gage. (*U.S. Forest Service*)

seed-bed preparation to two months after corn planting—coincides with a period of high erosion potential. Compared in this way, the corn fields near Lafayette would suffer more from erosion than those near Clemson. On the other hand, special care has to be taken at Clemson to protect the land from erosion during July and August. This points to an important application of the Wischmeier-Smith concept of erosion potential: the adjustment of the land use and management in such a way that the periods of soil exposure coincide as far as possible with the periods of least erosion potential.

**Measuring Precipitation.** In general two characteristics of precipitation are measured: the total amounts and the intensities. For the former purpose a container is needed that is provided with a horizontal opening of known area. In order to provide accurate information this opening should have a sharp edge, so that the collecting area is well defined. It is also necessary to protect the water in the container from evaporation until it has been measured. For this purpose a funnel with a small hole is built into the standard U.S. Weather Bureau rain gage. This funnel is removed for the period in which snowfall is expected, since snow could not pass through the hole. The cross-sectional area of the collecting vessel is one-tenth the catch area so that the amount of water can be readily determined by placing a scaled stick into it. The amount of water can also be found by weight.

Recording rain gages serve to determine rainfall intensity as well as the total amount. Several designs have been used. In most models the increasing weight of the water in the collector vessel is recorded continuously on a paper chart that is attached to a revolving drum. By changing gears such drums can be made to complete one revolution in 6 hours, a week, or almost any other time interval desired. For accurate determination of rainfall intensity the drum should turn fairly fast.

## *Infiltration and Percolation*

**Definitions.** *Infiltration* is the entering of rainwater into the soil. *Percolation* is the downward movement of liquid water within the soil. Sometimes the term "deep percolation" is used. It denotes percolation beyond the normal root zone of plants. The maximum rate at which water can enter the soil at any given moment is called infiltration capacity. Correspondingly, we also speak of percolation rate and percolation capacity. Infiltration and percolation are closely related. Infiltration provides the water for percolation; no water can percolate that has not infiltrated. On the other hand, the infiltration rate of a wet soil cannot exceed the percolation rate. For most purposes we can consider that infiltration is merely a special case of percolation.

If the water in the soil is not moving down vertically but in an approximately horizontal direction, it is called "lateral seepage." It results from ununiform permeability of the various soil layers Water that penetrates the upper part of the soil fairly quickly may be stopped by a slowly permeable layer and accumulate above it. The water flows on top of this layer until it again emerges on the soil surface farther down the slope, or the water may sink into deeper horizons if the impermeable layer is interrupted.

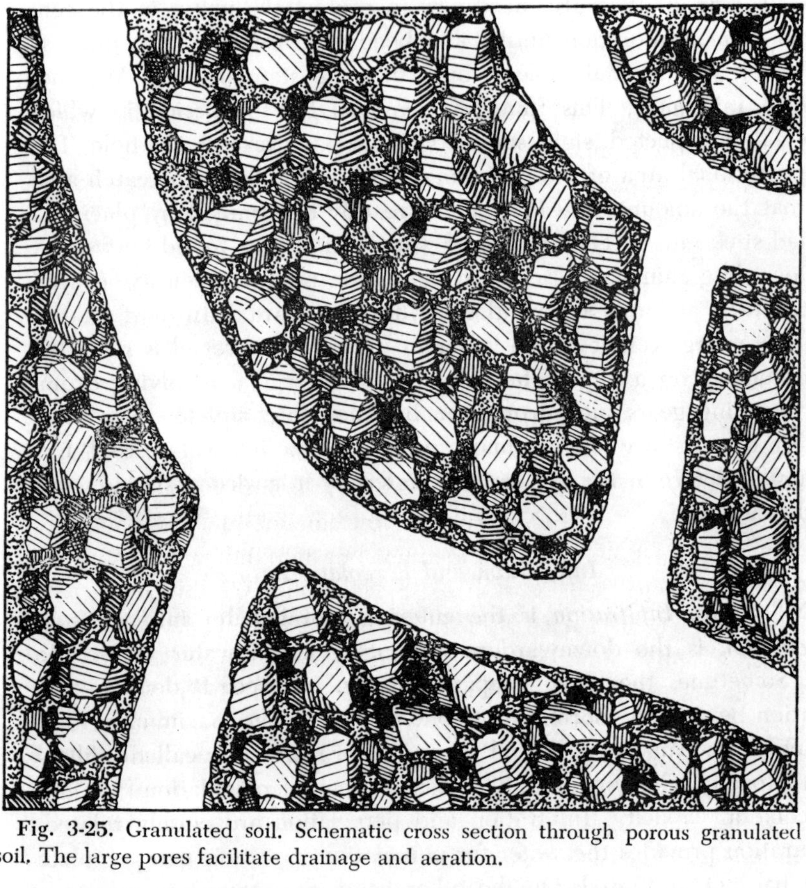

**Fig. 3-25.** Granulated soil. Schematic cross section through porous granulated soil. The large pores facilitate drainage and aeration.

**Factors Affecting Rates of Infiltration and Percolation.** Infiltration rates are determined by the infiltration-capacity rate and by the rate of water supply. As long as this is smaller than the infiltration capacity, the rate of infiltration equals the rainfall rate. As soon as the rainfall rate exceeds the infiltration capacity, runoff begins.

The soil characteristics that determine and limit infiltration capacity are structure, which is partly a consequence of texture, and moisture content.

In the consideration of infiltration and percolation the most important item of soil structure is pore size and the permanence of pores.

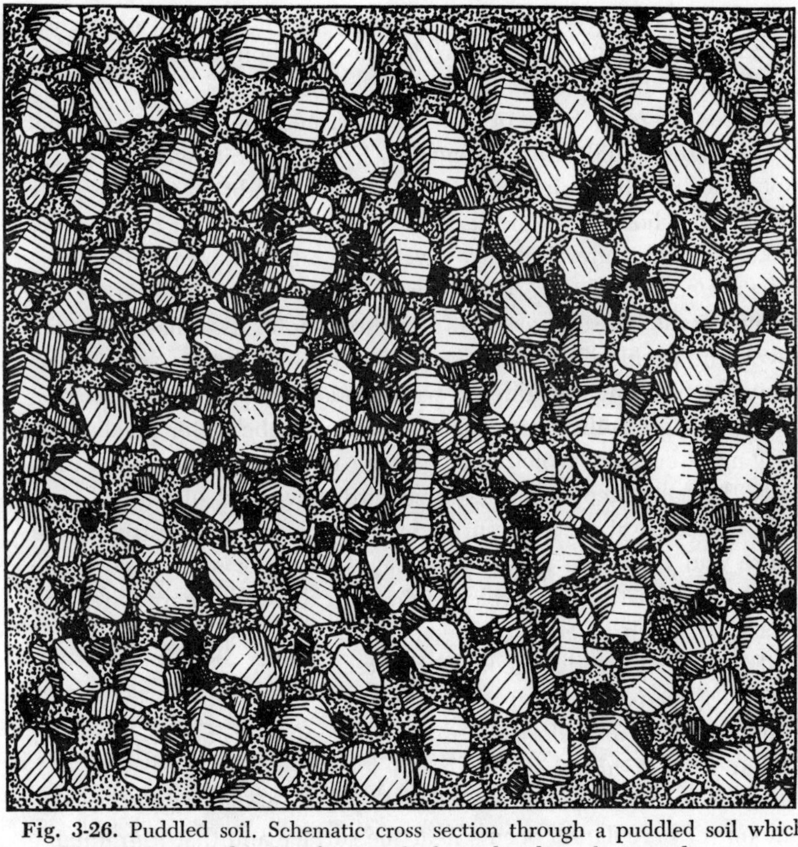

**Fig. 3-26.** Puddled soil. Schematic cross section through a puddled soil which has only small pores and is therefore poorly drained and poorly aerated.

*Pore Size.* The rate at which rain water can enter the soil is determined, to a large extent, by the size and the arrangement of the large pores. A soil with such pores is said to have "aeration porosity," because the pores have diameters large enough (0.06 millimeter and larger) to allow water to be drained out quickly and therefore serve to aerate the soil. They also allow air to escape from the soil to make room for the water. Formerly, soils with pores of this size were said to have "noncapillary porosity." Large pores may exist as a result of coarse texture or of aggregation.

*Permanence of Pores.* Infiltration rates can only be maintained if the original porosity remains undisturbed throughout the entire rainfall period. Soils that are readily dispersed will fill up the pore space, and the original infiltration rate is soon decreased. A soil of stable aggregation will maintain its infiltration capacity much better.

*Moisture Content.* Soils of low to intermediate moisture content have the highest infiltration capacities. Very dry soils without cracks do not absorb water quickly, since soil in this condition is not readily wetted. Soils whose pores are largely filled with water, of course, cannot take up much water and have a very low infiltration capacity.

*Soil Profile.* Thus far we have considered factors of infiltration and percolation as if the soil were uniform throughout. As a matter of fact, however, the profile characteristics of a soil are of outstanding importance in determining the rate of water intake. A pervious *A* horizon may have a high infiltration capacity only until the water reaches the underlying *B* horizon, which may be high in clay and low in aeration porosity. In cultivated soils the downward movement of water is frequently restricted at the very surface where the rain has destroyed soil structure and filled the pores with clay and silt or at the plow sole where the soil has been troweled tight by a dull plowshare when the soil was wet. Such layers have been called aptly the "bottlenecks" of the soil, since they restrict the flow of water.

A layer of coarse material with only very large pores under a fine-textured soil may also slow the downward movement of water until the soil above it is saturated.

Since soil serves as a reservoir of water, it is obvious that the greatest amount of percolation occurs near the surface and that it gradually diminishes farther down in the profile. In many storms only the upper few inches of the soil are wetted; in other words, percolation occurs only down to a few inches. This water is used up by evaporation and transpiration instead of traveling farther down into the profile. In many cases deep percolation occurs only in the winter and spring, while in the summer and fall there is no downward movement of liquid water in the lower soil horizons.

Temperature affects infiltration and percolation by influencing the viscosity of the water and by ice formation in the soil. The higher the temperature, the lower is the viscosity of the water and the greater is the rate of percolation. This effect is actually not very great, partly because it is compensated for by greater swelling of the soil colloids.

When a soil freezes that is only moist—not wet—the infiltration capacity is hardly diminished, because enough of the soil pores stay open. If another freezing period follows a rain and causes additional ice to be formed, the soil becomes increasingly less pervious to water, and eventually the infiltration capacity becomes zero.

*Table 3-5. Infiltration Capacities of Field Soils*

| Soil texture | Infiltration capacity, in. per hr |
|---|---|
| Loamy sand | 1–2 |
| Loam | 0.5–1 |
| Silt loam | 0.3–0.6 |
| Clay loam | 0.1–0.2 |

*Indirect Factors.* There are many ways in which infiltration and percolation are affected indirectly by affecting one of the previously named items. The amount of water available for infiltration is increased by depression storage created by cultivation, ridging, or contouring. Reducing evaporation serves the same purpose. Protecting the soil from the impact of rain helps maintain its aggregation and porosity. Holes put into the soil by mammals, worms, or insects serve to increase water intake capacity. Removing water from the soil by tile drainage, transpiration, and evaporation empties the large pores and frees the path for the intake of the next rain.

**Rates of Infiltration and Percolation.** Infiltration capacities from 0 to 60 inches per hour have been measured. Since infiltration capacity is usually greatest at the beginning of a rain and gradually decreases, it is best to determine the infiltration capacity after it has reached a fairly constant rate, if it is to be used for calculations of potential runoff. Infiltration capacities vary tremendously from case to case. The following table is given merely as a general indication of infiltration capacities for a number of textures of field soils. Infiltration capacities in protected woodlands are usually much higher.

Since these infiltration rates are for long-continued water intake, they depend on percolation capacities, and therefore the two may be considered equal.

**Measuring Infiltration and Percolation.** Infiltration rates can be determined under natural conditions by measuring rainfall, runoff, and estimating the other factors of the water cycle and by calculating the infiltration rate by "hydrograph analysis." The size of the area involved has to be known accurately. Either a natural watershed or a plot can be used. In both cases a runoff-measuring flume is required. Since occurrence of rain cannot be controlled, sprinklers can be used where comparisons of the effects of various methods of treatment on infiltration are desired. Such sprinkler-plot experiments are protected from the wind by means of tents [70]. Because of the large amount of water and the costly equipment required, infiltration studies are sometimes conducted on very small plots by means of so-called "infiltrometers" [29]. These apply water to the ground at exactly the rate of infiltration capacity. Hence only the amount of water used for given time intervals needs to be measured. Infiltrometers may be hand operated or automatic [56]. Usually the water is delivered without impact at the soil surface. Some infiltrometers are built with buffer compartments around the central plot in order to minimize water losses to the outside. Because of the small size of the plots, lateral movement of water in the soil can cause a fairly large bias in the results. Therefore they cannot readily be extrapolated to infiltration on a field scale.

For the measurement of percolation, "lysimeters" are used [41]. These consist of a container with an opening at the bottom. Soil is either filled into the container or the container is built around soil in its original structure. The water issuing from the opening at the bottom is measured manually or automatically. In order to obtain percolation conditions that approach the natural situation, the soil block should be undisturbed at least 6 feet deep, and provision should be made for runoff to occur so that no water will stand on top of the ground for a long period. Very few lysimeters are designed according to these specifications, and only the data from such lysimeters are of value for hydrologic studies. Some of the best ones are the watercycle lysimeters of the U.S. Department of Agriculture at Coshocton, Ohio [23, 30].

### *Runoff*

**Forms of Runoff.** Water draining off an area may take different courses.

*Surface Runoff.* Also called overland flow or sheet flow, this is the direct flow of water over the surface of the land. This is the most important erosive component of runoff, because it serves as transporting agent for detached soil particles from large areas.

*Subsurface Flow.* Water enters the soil but does not penetrate deeply because of the interference by a tight layer. The water travels with the slope of the land until it reappears as a "wet-weather spring."

*Ground-water Flow.* Water enters the soil and percolates down to the ground water. This in turn feeds the streams by springs or by lateral adjustment of the ground-water table.

*Stream Flow.* This is water flowing in a definite channel. Stream flow may be permanent or intermittent. It may be in a river bed or in a gully. There is no sharp boundary between stream flow and surface runoff. Stream flow may cause considerable erosion, but its nature restricts the damage done to relatively small areas.

**Characteristics of Surface Runoff.** The total amount of runoff, its rate of flow, its velocity, and its turbulence determine its erosive capacity. The product of the amount of runoff and its composition represents the amount of erosion.

*Amounts.* The amount of runoff may refer to a single storm or to a definite time period, such as the "water year." The water year adopted by the United States Geological Survey begins October 1 and ends September 30. In this way the beginning of the new year falls in a season when the rate of stream flow and the soil-moisture content are usually low, and therefore any errors in estimating these items will be small and will not greatly affect annual totals. Amounts of runoff can be expressed in inches for a given period, analogous to the recording of precipitation.

*Runoff Rates.* While there is no apparent close relationship between the total amount of runoff and erosion, a pronounced relationship usually exists between rate of runoff and erosion. During any one storm the runoff rate changes continuously and violently. On a watershed of a few acres the peak rate of runoff follows the peak rate of a runoff-producing rain by a few minutes. Since the high rates of runoff are the most erosive, we are particularly interested in determining the peak rates of runoff. Waterways, terraces, erosion-control structures, and the like are designed on the basis of peak rates of runoff. The average rate of runoff has very little meaning, since the same average rate may result from entirely different distributions of rates during the runoff periods. Toward the end of a storm and after it, the runoff gradually decreases and may continue for a long time at very low rates. During this "tail-out" period usually no erosion takes place. On the contrary, soil is deposited because of the reduced carrying capacity of the runoff water.

*Velocity.* Little accurate information exists concerning the velocity of overland flow. Outside of rills and gullies it is seldom more than ½ mile per hour (about ⅔ foot per second). If it were not for the rainfall splash that keeps the soil in constant motion, thin sheets of water at this low speed could not carry sand or silt for any important distance. Gully and other stream flow can be much faster, exceeding 10 miles per hour. Runoff velocity as a component of the erosive energy of water should receive serious attention.

*Turbulence.* Turbulence contributes much to the erosive action of a stream. When large amounts of water flow rapidly over uneven ground, the churning movement that results causes much erosive energy.

*Erosive Energy.* Runoff is both a detaching agent and a transporting agent. Its erosive energy is a function of its flow rate and its velocity. Sand and stones carried in suspension help in detaching soil but decrease the amount of additional soil that can be carried by the stream, as there is a maximum amount of soil load for a stream at a given flow rate and velocity.

*Runoff Composition.* Runoff water carries along with it suspended and dissolved material as well as bed load. If we are to determine the amount of loss of soil from a field, we have to multiply the amount of runoff by the concentration of the components. This can be expressed by the following equation:

$$\text{Inches of runoff} \times \frac{\text{lb of soil material}}{\text{acre-in. of runoff}} = \frac{\text{lb of soil material}}{\text{acre}}$$

Most erosion investigators determine the total amount of soil in the runoff. This is frequently done by taking an aliquot sample and evaporating it to dryness. In this way the material in solution is also included, and the resulting solid residue of the runoff is called "total solids." In case of runoff samples, medium to high in solids, "total solids" is practically identical with "soil." In nearly clear water the total solids include a considerable proportion of dissolved material.

This determination is simple and gives a picture of the loss of the soil body itself. From the point of view of the farmer and agronomist, however, the most serious immediate effect of soil erosion is the impairment of soil structure and the loss of available plant nutrients.

The same amount of total soil loss in two storms does not necessarily coincide with the same amount of fertility loss. Where erosion originates from a gullied field, subsoil and stones make up the bulk of the eroded material. On the other hand, sheet erosion from a productive, well-fertilized field may carry away an enormous amount of valuable plant nutrients.

*Phenomena Responsible for Bringing About Various Concentrations of Chemicals in Runoff Water.* If a hard, beating rain hits an unprotected field, the resulting runoff consists essentially of a suspension of soil in rain water. Since rain water carries usually only a low concentration of chemicals, such a runoff will differ little from a suspension of soil in distilled water. Normally the materials carried in surface runoff water are predominantly the finer and more valuable particles of the original soil. These carry with them microorganisms, exchangeable cations and absorbed phosphorus. If runoff occurs quickly after the start of a rain, soluble salts, particularly nitrates, accumulated in the surface during a dry period may be washed off before infiltration is able to carry them into the body of the soil.

Typical surface runoff water, assuming erosion occurs, is high in solid soil particles, especially clay and organic matter, high in total nitrogen, high in absorbed phosphorus, but low in soluble salts.

Percolation water contains a relatively high concentration of soluble salts but little or no organic matter, phosphorus, and colloids.

Subsurface seepage may be high in both soluble salts and colloids, especially organic colloids.

The composition of runoff changes greatly during any one storm [38]. Generally the first runoff is rather high in soil because of the removal of the soil particles detached by the raindrops. As the rate of runoff increases, the concentration of soil usually increases until just before the peak rate of runoff is reached, and then it gradually decreases until the last tail out of the runoff is practically devoid of soil.

In reporting the concentrations of plant nutrients in runoff it is important that only those parts of the individual elements be included that would become available to the plants in a reasonable period of time. It is difficult to determine this fraction in the case of nitrogen. Therefore total nitrogen is reported.

Phosphorus should include the organic, the exchangeable (with fluoride), and the dilute-acid soluble.

Of the bases, only the water-soluble and the exchangeable fraction should be considered as fertility loss. This includes the largest part of the organic portion of the bases. Most reports show a tremendous loss of potassium in runoff because the total potash in the runoff is determined. Soils have around 2 per cent total potassium but only about 0.01 per cent available potassium. The differences are not as great in the case of phosphorus, but they are also important.

Percolate water in areas of sandstone and shale has a considerably higher concentration of sulfate than surface runoff water. In calcareous regions the percolate water is rich in bicarbonate, while the surface runoff contains practically none. Determinations of sulfate and bicarbonate can therefore be used to estimate the proportion of surface runoff and percolate in stream water. This has some value in estimating the ground-water flow during periods of floods.

**Factors Affecting Runoff Characteristics.** Earlier in this chapter it was pointed out that rate and amount of runoff depend on many other components of the hydrologic cycle. Some of these again depend on other environmental factors. The outstanding ones are:

Rainfall—amount, rate, and distribution as to time and place
Temperature
Soil—types, condition, underlying strata, and topography
Size of drainage area
Plant cover—native or planted
Management—tillage and special water-management practices

The effects of these factors are so complex that, even though all of them are known, it is possible to calculate only very approximately the resulting runoff conditions. Wherever local conditions have been observed and studied for some time, more accurate predictions of the runoff conditions can be made for specific cases.

The larger the drainage area under consideration, the more will it be possible to predict the runoff characteristics from the rainfall records. This is the reason that flood forecasts for major streams are generally fairly reliable.

**Predicting Runoff Rates.** Even though an accurate prediction of rates and amounts of runoff is not possible because of the many factors that defy quantitative assessment, the necessity exists to estimate the maximum rate of runoff that can be expected to pass a given point. Without such information it would be impossible to determine adequate sizes of sod waterways, terraces, diversion ditches, culverts and bridges, and the level of road surfaces near creeks and rivers. Two methods are frequently used for this purpose. In either case it is first necessary to establish what degree of assurance is wanted that any runoff occurring in the future will be smaller than the calculated maximum rate. The absolute maximum runoff cannot be determined, because it is always possible that a storm of greater intensity and duration may occur than has ever been measured before. Moreover, it may not be economically sound to anticipate the largest known storm, because this might require excessively large structures.

Small storms are frequent. The larger the storm, the longer is the recurrence interval. It may be sufficient to design the structures so that they will handle runoff from all storms except the largest that might occur once in ten, twenty-five, or fifty years. Where precipitation data over many years exist, rainfall intensity-duration-frequency curves can be worked out that permit the statistical determination of the recurrence interval for storms of various duration and intensity [67, 74]. Generally ten-year return periods are used for calculating designs involving only earthwork and vegetation, while longer return periods (more severe maximum storms) are used in designing permanent structures.

The so-called "rational method" of predicting peak runoff rates takes into account the "time of concentration." This is the period required for surface runoff to flow from the most remote part of the watershed to its outlet. When a rainfall of a given intensity has lasted for this period, water from all parts of the watershed is reaching the outlet, and the runoff rate has reached its peak. The term "time of concentration" has only meaning for those cases where the rainfall rate is in excess of the infiltration capacity of the watershed. The equation used to calculate the peak runoff rate by the rational method is

$$Q = CiA$$

where $Q =$ the anticipated peak runoff rate for storms of a given recurrence interval, cu ft per sec

$C =$ the runoff coefficient, or the ratio of the peak runoff rate to the rainfall intensity (dimensionless)

$i =$ the rainfall intensity of the maximum storm expected for a given recurrence interval and a period equal to the time of concentration, in. per hr

$A =$ the watershed area, acres

The dimensions of this equation are cubic feet per second. They can also be expressed as acre-inch per hour since these two units are only 1 per cent apart. Actually, 1.01 cu ft per sec equals 1.00 acre-in. per hr.

The runoff coefficient varies, depending upon the perviousness of the soil, the slope, the vegetative cover, and the cultural practices. For land of high infiltration rates it is around 0.1, while it can rise above 0.8 for clay soils on hilly land.

In Cook's method the maximum runoff rate is estimated from the size of the watershed area, the characteristics of the watershed that affect runoff, the maximum storm of a standard recurrence period, the desired recurrence period, and the rainfall characteristics of the area. The equation used is

$$Q = PRF$$

where $Q =$ the maximum anticipated runoff rate at a certain point for storms of a given recurrence interval, cu ft per sec

$P =$ the maximum runoff rate from a watershed of given size and hydrologic characteristics (see below), assuming a ten-year recurrence rate and a rainfall factor of 1.0, cu ft per sec

$R =$ the geographic rainfall factor, dimensionless

In the United States the line representing the rainfall factor of 1.0 passes through New Jersey, along the Appalachian Mountains to northern Alabama, along the Mississippi River to central Iowa, and southwest through western Texas. The rainfall factors south and east of this line are higher than 1.0. This implies more intensive rains. North and west of this line the rainfall factors are smaller than 1.0.

$F$ is the recurrence-interval factor and is dimensionless. Since the basic runoff rate $P$ refers to a recurrence interval of ten years, the recurrence-interval factor for ten years is 1. For a longer interval it is larger (e.g., 1.4 for a fifty-year recurrence interval), and for shorter intervals it is smaller.

**Predicting Amounts of Runoff.** While it is desirable to know the peak rates of runoff to design waterways, weirs, terraces, and other

*Table 3-6. Maximum Runoff Based on Ten Years Frequency and Rainfall Factor of 1.0**

| Drainage area, acres | $W$ = summation of watershed characteristics† | | | |
|---|---|---|---|---|
| | 30 | 50 | 70 | 85 |
| | $P_{10}$, cu ft per sec | | | |
| 4 | 5 | 10 | 17 | 24 |
| 10 | 8 | 21 | 40 | 53 |
| 25 | 16 | 46 | 90 | 118 |
| 50 | 26 | 83 | 165 | 220 |
| 100 | 46 | 150 | 300 | 418 |
| 200 | 79 | 271 | 550 | 764 |
| 400 | 135 | 494 | 994 | 1390 |
| 600 | 185 | 675 | 1383 | |

* From *Farm Planners' Engineering Handbook for the Upper Mississippi Watershed*, U.S. Soil Conservation Service, Milwaukee, Wis., 1953.

† In order to arrive at the appropriate watershed-characteristics figure, the four properties of the watershed that affect runoff are considered (see Table 3-7), and their numerical values are added.

channels through which water has to flow, a knowledge of the maximum amounts of runoff water to be expected from a given drainage area is necessary to determine the volume of ponds and reservoirs that are to store this water. To arrive at this figure three items have to be known: the size of the watershed, the ultimate infiltration rate, and the rainfall characteristics of the area. The ultimate infiltration rate of a soil is reached when its entire profile has been saturated for some time and the infiltration rate is limited by the percolation rate of the

*Table 3-7. Classification of Runoff-producing Characteristics*

| Designation of watershed characteristics | Runoff-producing characteristics | | | |
|---|---|---|---|---|
| | 100 Extreme | 75 High | 50 Normal | 25 Low |
| Relief. . . . . . . . . . . | (40) Steep, rugged terrain; average slopes generally above 30% | (30) Hilly; average slopes of 10 to 30% | (20) Rolling; average slopes of 5 to 10% | (10) Relatively flat land; average slopes of 0 to 5% |
| Soil infiltration. . . . | (20) No effective soil cover; either rock or thin soil mantle of negligible infiltration capacity | (15) Slow to take up water; clay or other soil of low infiltration capacity, such as heavy gumbo | (10) Normal, deep loam; infiltration about equal to that of typical prairie soil | (5) High, deep sand or other soil that takes up water readily and rapidly |
| Vegetal cover. . . . . | (20) No effective plant cover; bare or very sparse cover | (15) Poor to fair; clean-cultivated crops or poor natural cover; less than 10% of drainage area under good cover | (10) Fair to good; about 50% of drainage area in good grassland, woodland, or equivalent cover; not more than 50% of area in clean-cultivated crops | (5) Good to excellent; about 90% of drainage area in good grassland, woodland, or equivalent cover |
| Surface storage. . . . | (20) Negligible; surface depressions few and shallow; drainage ways steep and small; no ponds or marshes | (15) Low, well-defined system of small drainage ways; no ponds or marshes | (10) Normal; considerable surface-depression storage; drainage system similar to that of typical prairie lands; lakes, ponds, and marshes less than 2% of drainage area | (5) High; surface-depression storage high; drainage system not sharply defined; large flood-plain storage or a large number of lakes, ponds, or marshes |

From *Farm Planners' Engineering Handbook for the Upper Mississippi Watershed*, U.S. Soil Conservation Service, Milwaukee, Wis., 1953. It is quite obvious that any of these methods can only yield approximations of the maximum expected runoff rates. Nevertheless, such information is indispensable for designs of many conservation practices.

least pervious horizon in the profile. The rainfall characteristics can be determined by using rainfall records of many years and preparing rainfall-intensity-frequency tables. In order to arrive at an estimate of the maximum expected volume of runoff the recurrence interval has to be established. In other words, it has to be decided whether the projected reservoir should be large enough to hold the runoff from the

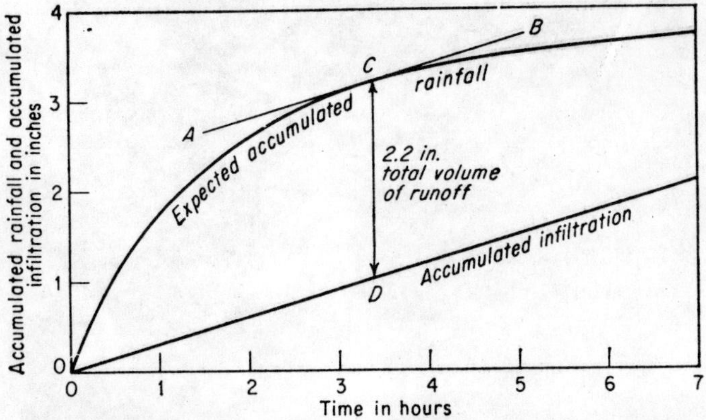


**Fig. 3-27.** Estimating runoff volume. The line *AB* is drawn parallel to the line representing accumulated infiltration as a tangent to the curve representing accumulated rainfall. This shows that on the left side of the line *CD* rainfall occurs at a rate greater than infiltration. Total runoff, therefore, is the excess of rainfall over infiltration; in this example 2.2 in. On the right side of the line *CD*, rainfall is slower than infiltration. Therefore no more water runs off. The total runoff—expressed in inches of depth—must be multiplied by the number of acres of the watershed in order to obtain the runoff volume in acre-inches.

largest storm that can be expected to occur once every ten or twenty-five or fifty years. Assuming it is planned to base the design on the largest storm in twenty-five years, a graph is plotted showing the maximum amounts of rain that can be expected to fall in increasing periods of time, as shown in Fig. 3-27. On the same graph the ultimate infiltration is plotted as accumulated infiltration, or "mass infiltration." Since it is assumed that the ultimate rate of infiltration for a given area is constant, accumulated infiltration appears as a straight line. On the other hand, the longer a rain lasts, the smaller is its expected maximum rate. Consequently, the slope of the line representing accumulated rainfall with time, or "mass rainfall," gradually decreases. At the time when the rainfall rate is the same as the infiltration rate, the

**Fig. 3-28.** Current meter. The cups above the torpedo-shaped weight will turn around their axis of suspension as soon as they get into flowing water. The flow is measured at two or more depths at regular intervals across the stream. The torpedo keeps the meter in position. (*USDA*)

maximum amount of runoff has been reached. After this, the rainfall rate is smaller than the infiltration rate, and no more runoff occurs. Therefore the total amount of rainfall minus the amount of infiltration to this time represents the total amount of runoff that can be expected. Figure 3-27 gives an example of using mass rainfall and mass infiltration to arrive at mass runoff.

This approach results in usable data, but it must be realized that several assumptions are made that are not necessarily correct. Infiltration is assumed to be constant. This is only so if the ground has been saturated just preceding the runoff-producing rainfall. On the other hand, it is assumed that the water that infiltrates does not contribute to runoff. If the rain lasts for several hours and slowly permeable layers exist in the subsoil, much of the infiltrated water will seep out farther down the slope and become surface runoff. It is also assumed

**Fig. 3-29.** Close-up of current meter. The velocity of the water is measured by electrically recording the number of revolutions of the conical cups on the meter. (*U.S. Geologic Survey*)

that interception storage and surface storage have been satisfied before the runoff-producing rain began.

**Measuring Runoff.** Runoff from natural watersheds is measured according to one of two principles. In one case the cross-sectional area of the stream is determined and the velocity of flow is measured at several depths at uniform distances across the stream. The depths selected are usually 0.2 and 0.8 of the total depth of water at the given point. The velocity of flow is measured by means of an electric-current meter. A number of conical cups are mounted around a vertical axis that is held in place by a torpedo-shaped sounding weight. These cups revolve under the action of the moving water, registering each or each fifth revolution by a clicking sound in earphones.

This method is employed in streams of medium and large size. By multiplying the flow rates in various parts of the cross section of the stream with the corresponding cross section, the discharge rate of the stream can be obtained.

The other principle is to allow the entire runoff water to pass through a weir or flume of specific design and to register the level of the water on the upstream side of this installation. Such a device has to be calibrated in order to determine the relationship between the height of the water level and the rate of water flow.

### *Interception and Surface Storage*

Interception is the interruption of the downward travel of rain or snow by vegetation or other soil cover. In the case of very dense vegetation all rain water is intercepted before it can hit the ground; therefore the energy with which it drops down is greatly diminished. The percentage interception (per cent of rainfall) varies with the density, the type, and the height of the vegetation. In a rainfall of long duration the amount of intercepted water that evaporates back into the atmosphere without striking the ground can be considerable. Some of the intercepted water glances off the vegetation and strikes the ground, some of it flows down the trunks of trees or stems of plants as "stem flow," and a portion remains in the vegetation after the rain stops and evaporates eventually. This latter amount of water is called "residual interception storage." The actual magnitude of residual interception storage varies greatly but can reach up to 0.2 inch for rain water and up to 0.5 inch for snow in dense coniferous forests [13].

Surface storage of water occurs as "depression storage" or as "surface detention." Depression storage represents rain water retained in depressions on the surface of the ground. The water cannot run off; hence it either infiltrates or evaporates. Surface detention is the transitory storage of water on the surface of the ground. This water is the layer just above the ground that is slowed down by friction or turbulence. Surface detention therefore is an expression of the slowing down of the surface runoff.

### *General Climatic Factors*

In addition to precipitation, other climatic factors have a pronounced effect on soil erosion by water. These factors are temperature, wind, humidity, seasonal and diurnal changes, and the land use that is adapted to the climate.

**General Type of Climate.** The various contributions of these factors make up the types of climate. In a classification of climates, sometimes the amount of moisture is stressed, sometimes the temperature, and sometimes a combination of these, as well as the nature of the precipitation.

Individual climates according to moisture are:

Superhumid
Humid
Subhumid
Semiarid
Arid

The effect of climate on natural vegetation and on soil development becomes apparent when world maps showing these three features are compared. They show that soil and vegetation boundaries coincide in a general way with climatic boundaries. Consequently, climate affects the erosion conditions of an area directly, as well as through the vegetation and the soils that occur. Water erosion and wind erosion must be considered separately when attempting to determine the effect of climates of different degrees of humidity on erosion. In arid climates wind erosion is more severe than in humid climates, while in the case of erosion by water this is reversed. However, the lack of organic matter in the soils of arid and semiarid regions makes these very susceptible to erosion by water as well as by wind. Another reason that water erosion in areas of low total rainfall can be serious is the fact that a large portion of the rain falls with high intensity during thunderstorms.

Tropical, temperate, and arctic represent the main grouping of climates according to temperature. Intermediate steps, such as subtropical, cold temperate, and subarctic are also recognized. Generally, rainfall intensity increases the closer the area is to the equator because of the ability of warm air to absorb much water. Therefore, erosion in tropical and subtropical countries is frequently quite serious. The amount of annual precipitation is affected by the prevailing winds, the distance from the sea, and the topography, as well as by the temperature. The cooler the climate, the more likely it is for rainfall to be gentle. As the winter becomes more pronounced, much of the precipitation occurs as snow. In arctic regions water erosion would be of little consequence even if there were extensive agriculture, because there is little rainfall.

The effective distance from the sea separates oceanic and continental climates. The ocean has a more uniform temperature than the land because the water reflects much of the heat of the sun, while soil and rock absorb this heat. Consequently, areas that are near enough to the sea have less temperature extremes than land in the middle of a continent. Where the prevailing winds occur from the land to the sea, however, it is possible for continental climate to exist very near the sea. The west coast of North Africa is an example of this.

From the viewpoint of erosion the energy of the falling rain is probably the most important single factor in the classification of climates. The American Middle West and Southeast are examples of high annual rainfall energy, while the gentle rainfall of England has a very low kinetic energy. Also, the distribution of rain and of rainfall energy is of profound effect upon the erosion hazard of a region. In some areas seasons with distinct dry and wet periods alternate.

Soil that has dried out thoroughly becomes highly erodible when exposed to intensive rainfall as it occurs in the monsoon belt. This is particularly true if little or no vegetation remains at the end of the dry period.

**Seasonal Distribution of Weather.** The change of the seasons brings about differences in soil moisture, plant growth, and soil protection as well as in precipitation and temperature. In some climatic regions intense rains fall predominately in certain seasons. In fact, in large areas no rain at all falls for several months each year. The outstanding effect of seasonal temperature differences is the change of rain to nonerosive snowfalls and the freezing of the soil surface that protects it from erosion.

**Indirect Effects of Climate on Erosion Hazards.** Climate is responsible for the development of the great soil groups of the world and consequently for their different erodibilities. Climate in its effect on soil moisture influences the erodibility of soil in two ways. Soils in dry climates are low in organic matter, while those that are well supplied with water usually have a higher organic-matter content and therefore are more resistant to erosion [33]. The climate also determines the type of crops that can grow in a certain area and consequently the type of agriculture practiced. In the same way, it determines the natural vegetation. Wherever this does not cover the ground completely, as is the case on arid and semiarid areas, the soils are exposed to the torrential rains that occasionally fall in such climates.

## *Topography*

Since "water seeks its own level," slope is a factor of outstanding importance in determining amounts and rates of both runoff and erosion. Six different properties of slope can be distinguished in this connection.

**Steepness of Slope.** The steepness, or gradient, of a land surface may be expressed in per cent or in degrees. With increasing steepness of slope the amount and rate of runoff increases, quickly at first, then the rate of increase decreases. On nearly level land there is little runoff, because much water is held in depression storage and because the gradient that causes flow is small. As the slope increases, depression storage decreases. Eventually an additional increase in steepness does not increase greatly the amount of runoff, since surface storage is very small on steep slopes and the infiltration rate is not directly affected by steepness.

The increasing velocity of runoff that results from increasing steepness of slope makes the water a better transporting agent and allows the raindrops to hit the ground more directly, since the sheet of water on the surface becomes thinner as the flow velocity increases. Experiments have shown that by and large, the amount of erosion per unit area increases 2.5 times as the degree of slope is doubled.

It must be realized, however, that soil changes normally go along with slope changes. The steeper the slope, the greater has been the removal of clay from the surface and the higher is the percolation capacity of the soil, because the accumulation of clay in the *B* horizon becomes less, until eventually on very steep slopes there is no *B* horizon. Consequently, the effect of slope as such is somewhat mitigated by this concomitant change in soil profile.

**Length of Slope.** The effective length of slope in the study of erosion is measured from the beginning of overland flow to the point where the water flows into a defined channel or where the slope diminishes so much that deposition occurs.

During runoff, water accumulates as it flows down a slope. Consequently, more water flows over the lower part of the slope, and it flows faster than it does over the upper part of the slope. The total amount of runoff per area may not be greatly affected by the length of the slope. As a broad average, soil loss increases 1.5 times per unit area when the slope length is doubled [2, 31, 48, 59, 61].

**Configuration.** Slopes may be convex (bulging) or concave (hollow). It is a general observation that sheet erosion is more severe on convex slopes than on concave slopes in the same field. There are a number of reasons for this:

1. Soil on a convex slope dries out readily and therefore contains less humus than on a concave slope. Dry soil breaks up more readily when it is wetted by rain than moist soil—as it would be on the concave slope—and the low humus content makes its aggregation less stable.

2. Since water generally flows away from a convex slope, it removes humus and clay without replacement from higher areas. Consequently, it becomes poorer in these components and more detachable.

3. On convex slopes water spreads out and does not form a thick sheet that acts as a protection against the impact of raindrops as it does on concave slopes.

4. Splash transport is larger on convex than on concave slopes.

5. On a convex slope the steepness increases toward the bottom of the hill. Therefore the runoff velocity increases as it flows down the hill and thus is able to carry more and more soil. On a concave slope the situation is reversed, and soil is dropped out as the velocity of runoff decreases.

Obviously, water tends to concentrate in concave slopes, and the runoff itself may act as detaching as well as transporting agent, thus forming gullies.

**Variation in Steepness.** Slopes are seldom uniform from the crest to the valley. Where this variation is pronounced and where steeper and flatter slopes alternate, it may have a considerable effect on runoff and erosion. Such differences in steepness usually go together with changes in soil characteristics, causing soil of high and low infiltration capacity to alternate. No research on these effects has been published, but it is likely that areas of variable slope will have less erosion than those of the same average slope that are more uniform. The flatter concave areas have the tendency to slow down the runoff and to settle out the soil. An indirect effect of variation of steepness of slope on erosion is that the farmer uses less row crops on such areas, because cultivation is more difficult.

**Aspect of Slope.** Experience shows that slopes that face south and west in the Northern Hemisphere suffer more from erosion than those that face north or east. South slopes are exposed to much greater variations in temperature and moisture than slopes facing in other directions. Southern slopes face the sun more directly, so that its energy heats up the ground and dries it out much more than level land. Consequently, soils of southern slopes are lower in organic-matter content, and when they are very dry, they are much easier detached than somewhat more moist soils of northern exposure. During the winter south-facing soils are subject to the structure-destroying effect of freezing and thawing much more frequently than north-facing soils.

Because of the more favorable conditions for plant growth, north-facing soils are frequently better covered with vegetation than south-facing soils. Evaporation is greater on south-facing slopes than on north-facing slopes, and the steeper the slope, the greater is this difference. The following findings by Eser [22] illustrate this fact. For

*Table 3-8. Relative Evaporation from Soil*

| Slope | Facing south | Facing east | Facing west | Facing north |
|---|---|---|---|---|
| 15° | 100 | 86.3 | 84.1 | 71.0 |
| 30° | 100 | 80.6 | 73.1 | 52.7 |

this reason the amount of runoff has been found to be smaller from the south-facing slopes, but the amount of erosion is larger on south-facing slopes for the reasons given. This is exemplified by the research by Wollny [1].

In areas where most storms occur when the wind comes from a certain direction, soils with this aspect should have relatively higher erosion than they ordinarily would. No research on this item has been found in the literature.

**Microtopography.** By *microtopography* is meant the unevenness of the soil surface. This is generally of so small a dimension that it is not shown on any topographic map, no matter how large the scale.

Minor depressions and elevations as those caused by tillage, stepping of animals, accumulations of soil at fence rows, and gullies are part of microtopography.

*Table 3-9. Effect of Aspect on Erosion*
(Erosion in tons per acre per year)

| | Slope | Facing north | Facing south |
|---|---|---|---|
| 1882 | 15° | 4.5 | 6.3 |
| | 30° | 8.8 | 10.5 |
| 1883 | 15° | 2.1 | 2.6 |
| | 30° | 4.6 | 4.9 |

This unevenness of the soil surface may form many small depressions, yielding storage capacity for water and thus giving it more time to infiltrate. Or it may take the form of dead furrows, gullies, wagon tracks, or cattle paths that concentrate the runoff water, thus facilitating and speeding up runoff velocity and enhancing erosion. Contouring and terracing are examples of man's attempt to make use of the beneficial properties of microtopography.

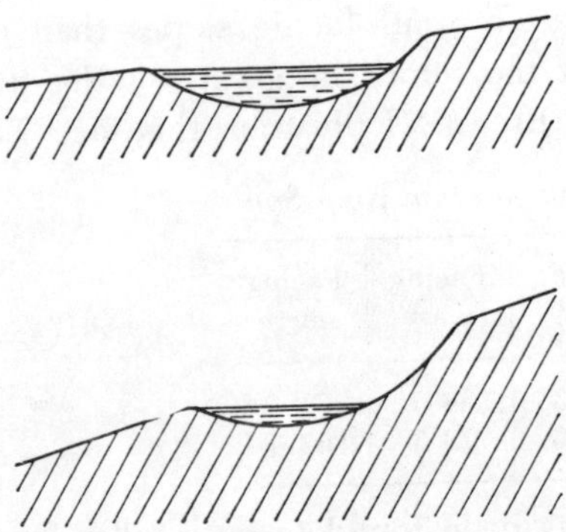

**Fig. 3-30.** Microtopography and slope. A depression in the soil of a 10 per cent slope (top) can hold much more water than an identical one on a 30 per cent slope (bottom).

The effect of microtopography on runoff and erosion changes with the steepness of the slope. It is most effective in reducing runoff and erosion on slopes of intermediate steepness, from about 2 to 7 per cent. On flat land the runoff gradient is so small that runoff velocity, and with it erosion, is small. On steeper land small depressions in the soil cannot retain much water, since the downhill rim of each hole is not much higher than its lowest point. On slopes of over 20 per cent gradient microtopography has very little effect on the amount of erosion.

**Size of Watersheds.** The size of the watershed affects the rate of erosion only indirectly, by determining the amount and velocity of runoff that passes over a given point. This has been discussed previously.

## *Soil Properties*

In order to identify the soil properties that affect erosion by water, one must consider the components of erosion, detachment and transportation. Soils of high detachability and high transportability are highly erodible. The effect of soil conditions on rainfall disposal should be considered. The infiltration capacity of the soil is one of the determining factors of runoff rates and amounts and therefore affects indirectly its erosion hazard.

The soil factors affecting infiltration capacity have been discussed previously. Only those affecting detachability and transportability will be taken up here.

**Factors Affecting Detachability.** To protect a soil particle from being separated from the soil mass—in other words, to reduce its detachability—it has to be tied together with enough other particles that the energy of the detaching agent cannot move it. This means that the soil has to be aggregated. The necessary size and weight of these aggregates depend on the amount of energy of the rain or runoff that might cause the detachment. The greater the energy, the larger the aggregate has to be to withstand the impact. Clay is the main cement of the soil aggregates. Its fine particles fit into the openings between sand and silt and create a soil of a much greater consistence throughout the many surface contacts than a soil that is made up predominantly of silt and sand [4]. It is the clays of high specific surface (montmorillonites and related clays) that are most efficient in holding soil particles together. Whether this consistence will be maintained when the soils get wet depends on the nature of the clay, on the exchangeable cations, on the presence of stabilizing organic compounds, and on the rate of wetting. Clays that are largely saturated with calcium and magnesium are generally flocculated and are of more value in stabilizing soil aggregates than clays that are predominantly sodium saturated and consequently dispersed [50]. Clays with expanding crystal lattice swell greatly on wetting and therefore have the tendency to break up soil aggregates; on the other hand, they enter into complexes with organic compounds quite readily, and it is this clay–organic-matter combination that furnishes stability to the soil [55]. If the wetting is slow, so that air can be expelled from the soil through its natural fissures, the aggregates are much more likely to be maintained than if water enters from all sides, engulfing the soil air. This air is then compressed by the capillary attraction of the water toward the middle of the aggregate and breaks up the soil to supply an avenue of escape. The stability of the soil aggregates varies with the moisture content. Intermediate moisture is the most favorable. If a soil is moist at the beginning of a storm, not much air is contained within the pores, and the aggregates have a good chance to survive. Dry soil suddenly wetted breaks up much more readily, because the air on the inside is compressed and forces its way out through the wetted layers. If the soil has been wet for some time preceding the storm, some of the organic binding substances have been dissolved, and the water has separated the individual clay, silt, and sand particles, thus weakening their attraction to each other. Consequently, both a wet and a dry soil are more prone to erode than one of intermediate moisture content.

Much more research is needed to provide a full picture of the role that the various organic compounds play in forming complexes with clays. It is well known, however, that water-stable soil aggregates can only be formed in the presence of both clay and organic matter. Since this organic matter is constantly oxidized in the soil, especially in a well-drained, cultivated soil, it is necessary that a new supply of crop residues is available at all times for the replenishment of these vital compounds [60]. Organic matter helps to maintain soil aggregates in a variety of ways: in the form of the clay-organic complexes that have just been discussed, as microbial mycelia and mucus, as root hairs that form a firm contact with the soil particles, and as the excreta of earthworms and microanimals. In the form of plant remains, organic matter can serve as a mulch that protects the aggregates from the direct impact of the raindrops.

The plant-nutrient status has only an indirect effect on the detachability. Fertile soils support much plant growth, which provides protection against the impact of rain and a source of organic matter and thus encourage microbiologic activity.

**Factors Affecting Transportability.** *Size of Detached Particles.* The settling velocity of a solid particle in a liquid is proportional to the square of its diameter. Its transportability increases as its settling velocity decreases. It is obvious, therefore, that the smallest detached particles are the most readily transported. Sand particles or soil aggregates of sand size or larger are only transported by rapidly flowing water or by water that is constantly agitated by rain splash. Dispersed clay particles stay in suspension almost indefinitely.

*Particle Density.* The true particle density of the mineral components of soils varies within fairly narrow limits, around 2.65 grams per cubic centimeter. The density of organic matter is about one-half that amount. Organic matter is therefore much more readily floated off the field than mineral soil grains. But when soils are aggregated—and most soils are—the crumbs are made up of mineral and organic particles and pore space. Whether this pore space is filled with air or water, it helps to lower the bulk density (average specific gravity) of the aggregates. Therefore an aggregate can be transported more readily than a sand grain of the same size. Normally, of course, aggregated soils resist transportation more than dispersed soils, because the aggregates are bigger than the individual particles.

**Methods of Estimating Erodibility of Soils.** Since detachability, transportability, and infiltration capacity combine to make up the erodibility of a soil, each of these three factors has to be studied.

Detachability may be determined in the field by the use of splash boards or in the laboratory by artificial rain on soil samples in cylinders [20]. In both cases the amount of soil splashed from a given area per unit time is measured.

Transportability is estimated by the determination of the proportion of soil particles of clay or silt size that can be suspended in water by gentle shaking. The opposite of this, the percentage of water-stable aggregates, can also be determined for this purpose.

The infiltration capacity can either be determined directly on natural watersheds or small plots or be inferred from a determination of the aeration porosity.

**Interrelation with Other Factors.** No determination of soil characteristics will make it possible to line up a large number of soils in the sequence of their erodibility, because other factors will affect this sequence.

*Rainfall Intensity.* A gentle rain may be sufficiently intense to detach much of a certain sandy soil, but the infiltration capacity may be large enough to prevent runoff, and therefore the rain causes no erosion. On a silty clay loam the same rain may cause much runoff and sufficient detachment to bring about appreciable erosion. If the rain is more intense, runoff may result on the sandy soil that will carry off much of the detached soil. A much greater rate of runoff would probably result on silty clay loam, but its detachability is so low that the rate of erosion is not very greatly increased.

*Slope.* The order of the erodibilities of two soils on one slope may be opposite to that on a different slope. On a slight slope sand may erode less than clay; on a steep slope this relationship may reverse. The case is parallel to the effect of variations in rainfall intensities on erosion from different soil. An increase in slope does not increase the erosion hazard of all soils at the same rate. This is because the effects of slope variation on detachment and on transportation are entirely different. An experiment by Duley and Hays [16] illustrates this situation. The relative erodibility of these two soils is reversed by changing the slope of the land. This shows the difficulty of classifying soils as to their erodibility on the basis of physical characteristics, for instance, the erosion ratio and the dispersion ratio as suggested by Middleton et al. [50, 51].

*Table 3-10. Effect of Slope on Detachment and Transportation*

| Soil type | Lb of water required to remove 1 lb of soil | | Soil characteristics | | |
|---|---|---|---|---|---|
| | 8% slope | 16% slope | Infiltration capacity | Detachment hazard | Transportation hazard |
| Sandy loam | 179 | 7 | High | High | Low |
| Silty clay loam | 65 | 24 | Low | Low | High |

*Agriculture.* Land use, cropping, and tillage practices are determined to a large extent by the nature of the soil. Therefore under practical conditions of agriculture a soil of high physical erosion hazard may actually not erode much, because land use and treatment may give it considerable protection. On the other hand, intensive cropping and cultivating may increase the erosion hazard of a normally fairly erosion-resistant soil.

**Summary.** A summary of the effect of soil properties on soil erosion by water is listed below:

SOIL PROPERTIES THAT REDUCE DETACHABILITY
(*Properties that make for soil stability*)

High active organic matter content
High clay content
Prevalence of divalent ions among the exchangeable cations
High content of water-stable aggregates
High amount of microbial activity
High fertility to stimulate microbes and crops
Intermediate moisture content at the beginning of the storm
Consolidated surface

SOIL PROPERTIES THAT REDUCE TRANSPORTABILITY
(*Properties that make for large particles*)

High percentage of large primary (sand and gravel) and secondary particles (water-stable aggregates)

High percentage of organic matter (stabilizes the aggregates, but at the same time decreases their average density)

SOIL PROPERTIES THAT REDUCE RUNOFF HAZARD
(*Properties that make for high infiltration capacity*)

High percentage of large primary and secondary particles
High percentage of large continuous pores (large aeration porosity)
Intermediate moisture content at the beginning of the rain
High percolation capacity (absence of impervious strata)

Expressing the effects of texture on the erodibility of soils in broadest generalizations, the following can be said:

Sandy soils are readily detachable but not readily transportable; moreover, they have high infiltration capacities. The hazard of erosion by water is therefore small, except on steep slopes.

Soils of medium to high clay content have low infiltration capacities, and they are readily transported by water after they are dispersed (detached), but since their detachability is generally low, they are fairly resistant to erosion.

### *Surface Cover, Land Use, and Cultural Practices*

In the study of accelerated—man-induced—erosion, a thorough analysis of the effects of the use and treatment of the land must necessarily take an important position, inasmuch as these factors affect both the surface cover by plants and plant residues and the condition of the soil. Just as important is a knowledge of the natural vegetation and its features affecting runoff and erosion, since this represents the base line from which accelerated erosion is measured.

**Plants and Plant Residues Covering the Ground.** Plants affect runoff and erosion conditions in a variety of ways. Protection from the impact of rain is provided by the plants as well as by their residues on the ground. The extent of such protection depends on the size and growth habit of the plants. Generally, the more prostrate the plant, the more erosion control it provides. Tall and erect plants are of little value in this respect unless their residues cover the ground. The size and shape of the leaves determine the size of the water drops that drip off the plants. Rain hitting an alfalfa field will be broken up into small, relatively harmless droplets, while water dripping from corn falls in much larger drops.

Transpiration removes much water from the soil, thus making room for intake of rain water. This reduces the amount of water that would otherwise run off. Residues and the stems of plants furnish resistance to overland flow, slowing down runoff velocity and reducing erosion.

On the other hand, a dense vegetation protects the ground from excessive evaporation, thus keeping the surface moist longer than a bare soil and making the soil aggregates less liable to detachment.

The root systems of plants are very different in their growth patterns and their effects on the soil. The fibrous roots of most grasses are very effective in forming stable soil aggregates. Deep taproots of some legumes and of many other plants open up the soil to considerable depth. The few, relatively thick roots of the soybeans and, still more, the tubers of the potatoes fluff up the soil and make it erodible.

Summarizing, we can say that the presence of a dense stand of plants, especially if plant residues remain on the soil, is a very excellent way to reduce the detachment hazard and the transportation hazard, to increase infiltration capacity, and as a consequence, to decrease runoff and erosion.

**Cropping Practices.** Cropping is the fundamental reason for acceleration of erosion over geologic norms. The need to change the environment so that it is favorable for one type of plant and the removal of the others exposes the soil and leaves it loose and liable to erode.

*Tillage* in itself is the most erosion-inducing factor in agriculture. Usually the more intensive the tillage, the more it induces erosion. Tillage increases the detachment hazard, it actually detaches soil, and it tends to oxidize its organic matter. This decreases aggregation and reduces the infiltration capacity. Plowing creates a plow pan. The weight of tillage implements and other agricultural machinery compresses the soil, reducing its amount of large-pore space and consequently its infiltration capacity. All this results in higher runoff and erosion rates.

Of course, it must be recognized that in many respects tillage works in the opposite way. Loosening the soil with the plow or any other implement makes it temporarily more receptive to rain water. Mulch tillage, contouring, and subsoiling are tillage methods of more lasting benefit [3].

*Planting methods* vary in their effect on erosion, depending on the species, the stand density, the distance between the rows, and the direction of the rows with respect to the slope. The denser and the more nearly on the contour the planting is made, the less erosion will result.

*Fertilization* helps to ensure stand and causes faster and heavier growth and is consequently a help in protecting the soil and in creating more residues. Manuring can serve both as a fertilizer and a ground cover.

*Weeds,* like all other plants, protect the soil from erosion. Combating them by cultivation increases the erosion hazard.

*Harvest methods* vary greatly in their effects on soil erosion by water. They are enumerated here in the order of decreasing erosiveness.

1. Removal of both tops and roots. Examples of this are potatoes, where the vines are used for bedding or are burned. In some parts of

the Orient the entire rice plant is harvested; grain, straw, and roots are removed. The latter are used as fuel.

*Table 3-11. Relative Efficiency of Crop Cover to Protect Soil from Erosion*
(In order of decreasing efficiency)

| Land-use groups | Examples |
|---|---|
| Permanent vegetation | Protected woodland |
| | Prairie |
| | Permanent pasture |
| | Sodded orchard |
| | Permanent meadow (slough) |
| Grass-legume meadows | Alfalfa—brome grass |
| | Clover—timothy |
| | Alta fescue—birdsfoot trefoil |
| Small-seeded legumes (without grass) | White clover |
| | Red clover |
| | Alfalfa |
| | Sweet clover |
| | Crimson clover |
| Small grains | Rye |
| | Wheat |
| | Barley |
| | Oats |
| Large-seeded legumes | Soybeans |
| | Cowpeas |
| | Peanuts |
| | Field peas |
| Row crops | Cotton |
| | Potatoes |
| | Tobacco |
| | Soybeans |
| | Corn |
| Fallow | Summer fallow |
| | Period between plowing and growth of crop |

Special cases: orchards (type of cultivation), gardens, truck crops, grazed woodlands, overgrazed pastures, abandoned land, weed patches.

2. Removal of entire tops or of roots. Some of the following belong in this group: hay harvest, fodder or silage, hemp, and the root crops, such as potatoes, sweet potatoes, beets, and peanuts.

3. Removal of parts of tops only. This is the case with most nonforage crops. Different parts are removed, e.g.; grain from small grains, corn, beans, seed clover, lint from cotton, fruit from fruit trees, leaves

from tea and tobacco, and stems from lumber trees. Since much of the crops remain in the field, the depletion of organic matter and minerals is not quite as serious as in the previously mentioned cases.

4. Pasturing. The crop is gradually removed, and part of the organic matter and the minerals is replaced by the manure. If pasturing is managed carefully, the amount of erosion is negligible, but overgrazing, pasturing when the soil is wet, or allowing the animals to make paths may make pasturing a very erosive form of harvesting.

5. Special cropping practices. Detasseling corn by means of moving derricks ("highboys") carrying five to eight people is bound to compact and puddle the soil in any weather. Irrigation may cause erosion if water is allowed to flood sloping land or sprinkler drops are too large and the irrigation rates too high.

**Relative Erosiveness of Crops.** Since the cropping practices as well as the nature of the crop plants determine the erosion hazard, it is impossible to rank the erosiveness of different land uses and crops with any accuracy. In Table 3-11 an approximate order of erosiveness of the crops is given.

**Effects of Rotations on Soil Conditions and Erosion.** The continuous improvement of power equipment and the availability of relatively inexpensive fertilizer, especially nitrogen fertilizer, have changed the effects that individual crops may have on succeeding crops and on erosion. This development is far from finished. New cultural practices are introduced every year. An attempt to rank rotations as to their effects on soil conditions and erosion presupposes a definite point of view in the acceptance of cultural practices. An added difficulty is that, of necessity, there is no research information on the more modern cropping methods that has lasted long enough to show the eventual effects of their repeated use.

More specifically, some of the problems are these: To what extent can nitrogen fertilizer replace the beneficial action of legumes? Are row crops "erosive" and "soil depleting," if they are amply fertilized, mulch tillage is used, and the weeds are controlled with herbicides? Will interplanting of grasses and legumes between corn rows make a rotation unnecessary on many farms of the Middle West?

While no clear answers exist to these and other similar questions, some basic facts permit a ranking of rotations based on the experience of the last decade. Meadows, especially those with a large proportion of grass, have a tendency to improve the soil structure because of the fibrous and dense root systems of grasses and because they provide a continuous cover for the soil. Row crops that are planted into bare soil and that are mechanically cultivated are the most erosive crops, especially where the time of exposure of loose soil coincides with the season of the most severe storms, as is the case in the Middle West. Small grains are intermediate in their effects on erosion. These facts are illustrated in the table on relative erosion losses developed by the U.S. Department of Agriculture.

Recognizing the many limitations that variations in cropping practices, climate, soil, and topography impose, the following ranking of rotations is merely suggested:

SAMPLES OF CROP ROTATIONS IN SEQUENCE OF INCREASING EROSION HAZARD [57]

1. *g-M-M-M*
2. *R-g-M-M*
3. *R-g-M*
4. *R-R-g-M-M*
5. *R-R-g-M*
6. *R-R-g-M-R-*$g_{(sc)}$
7. *R-*$g_{(sc)}$
8. *R-R-R-g-M*
9. *R-R-*$g_{(sc)}$
10. Continuous row crop

where *R* = row crop (corn, soybeans, etc.)
*g* = small grain (wheat, oats, rye, etc.)
*M* = meadow
(*sc*) = sweet-clover intercrop

**Value of Pasture and Forest for Soil Conservation.** The fact that the ground is permanently covered by vegetation and mulch makes pasture and forest some of the most soil-conserving forms of land use. Which of the two will be most efficient in protecting the soil from erosion depends largely upon the methods of management. In a forest the impact of the rain is broken by the various stories of branches and finally by the leaf litter on the ground. Thus the water reaches the soil practically without kinetic energy. It readily infiltrates into the leaf mold ($A_0$ horizon) and the porous, highly organic topsoil ($A_1$ horizon). Whatever water does not percolate farther down into the soil profile is carried safely in channels created by root decay or in rodent runs down the slope into a draw or ravine. This situation changes if burning destroys the leaf litter or even the trees, if pasturing animals compress the soil and destroy the channels near the surface, and if unwise logging methods cut gashes into the soil in snaking out the trees and create potential gullies.

A pasture does not have the same advantage of breaking the impact of the rainfall by several layers of tree branches, but its dense

**Fig. 3-31.** Overgrazed mountain range. Severe erosion was caused by runoff from one rain on this overgrazed range. Soil, rock, and debris from this area created heavy damage to meadows and irrigation ditches below. (*USDA*)

cover of grasses, legumes, and other low-growing plants is just as effective in intercepting the raindrops. In fact, where the sward is continuous, it provides better protection against erosion in the draws than leaf litter in the forest, since it cannot be swept away by the water. The tramping by livestock causes pasture soil to be dense, but the infiltration capacity is nevertheless fairly high, probably because of the many roots that leave channels when they die.

The amount of erosion from a well-managed pasture is negligible, but an overgrazed pasture can erode as rapidly as a cultivated field. If the grazing animals eat the forage down to the ground, there is little, if any, interception, and the root systems of the plants are weakened because of a restricted plant surface for photosynthesis. As plants die, open spots develop in the pasture, and erosion increases. While cattle create this condition, sheep can cause more harm to the land because they eat the plants closer to the ground. Goats do even greater damage by eating also the woody part of the vegetation, which

**Fig. 3-32.** Overgrazing causes erosion. On the left side of the fence, overgrazing has resulted in severe erosion. On the right, moderate grazing has maintained a nice blanket of vegetation to control erosion. (*State Information Office, Pretoria, South Africa*)

they can digest better than other animals. Examples of overgrazed, badly eroding pastures can be found in the United States as well as in many other parts of the world.

## MEASURING SOIL EROSION BY WATER

Various approaches can be used in measuring soil erosion. Some methods serve to measure accumulated erosion for a long time period; others measure the erosion of an individual storm. It is also possible to measure detachment alone. No methods have been proposed to measure transportation independently of detachment under fully natural conditions. It can be done where a controlled amount of soil is added to runoff water.

### *Measuring Detachment*

Detachment by rainfall splash can be measured in the field with vertical splash boards [20, 52] or collector vessels placed into the ground so that the rim protrudes only slightly above the soil, or it may be measured in a laboratory.

### *Measuring Erosion of an Individual Storm*

The size and shape of the area from which erosion is measured determine largely what type of information is obtained.

**Small Plots.** Small—usually rectangular—plots are used if the results are to be referred to one soil type and one slope gradient. Many of these plots are so small that the entire runoff of an individual storm

**Fig. 3-33.** How losses are measured. Runoff is concentrated at the base of an enclosed plot by a trough. It is subdivided in small flumes and an aliquot is collected in a tank. (*USDA*)

can be collected in a tank at the bottom of the plot [2, 31, 37, 44, 59]. This facilitates accurate determination of the soil lost. In other cases dividers are used to sample only a fraction of the runoff. The disadvantage of the plot method is that natural overland flow from neighboring land is eliminated, and sometimes the runoff in the plot area concentrates on one side, flowing along the divider strips. This may result in gullying. After several severe storms erosion has lowered the soil surface on most of the plot, but the overflow plate at the bottom has maintained the elevation of the soil in the lower part of the plot. Movable overflow plates are used, but each time the overflow plate is lowered the first storm will cause an excessive amount of erosion, since the support of this soil is removed. When the plots are only a few feet wide, as most of them are, cultural operations have to be performed with hand tools. Consequently, these soil conditions are not identical with those under ordinary farming.

**Watersheds.** Measurement of erosion from watersheds avoids some of the shortcomings of the use of small runoff plots, but it introduces other problems. A watershed is the natural unit of runoff. It is defined as a drainage area whose runoff flows past one point. There is no size limit to a watershed as such, but for use in erosion studies the area has to be small enough to carry most of the soil that moves downhill past the outlet point of the watershed with the water during a rain. The maximum size depends on the topography. The steeper the land, the larger can be the watershed. In flatter areas, especially where the slope gradient decreases downstream to form alluvium (bottomland), the watershed should be so small that the outlet is above the alluvium; otherwise, too big a part of the eroded soil would be deposited before it reaches this point. Of course, the best size of a watershed for erosion determinations depends on the specific purpose of the study. Where the effects of different types of land use or treatment are to be investigated, watersheds should be less than 20 acres in size. In these small areas most of the soil that erodes during a heavy storm reaches the outlet. Most single land-use watersheds studied are actually less than 10 acres in size.

The advantage of using a watershed is that both the cultural practices and the hydrologic behavior are identical to those existing on large areas of land. On the other hand, most watersheds contain several soil types and erosion classes and a variety of slope gradients, lengths, and aspects. Consequently, the results cannot as readily be referred to specific soil and slope conditions as those from plots.

Since the total amount of runoff water from a natural watershed, even though the area is only an acre in size, is too great to place in a tank, a runoff measuring device and a runoff sampler are used. The runoff is normally measured in a flume equipped with an automatic water-stage recorder. The flume is calibrated, so that the rate of runoff for every height of water passing through it is known. Various devices have been developed to obtain aliquot samples of the runoff passing through the flume. Examples are the "sludge box," for the deposition of the coarse fraction of the eroded material, and the "silt sampler," for the collection of a portion of the runoff water containing soil in suspension [2, 24, 31, 53, 59]. Other devices are the Indiana aliquot runoff sampler [39] and the "Coshocton wheel" [54].

Usually, the total amount of soil loss is determined. When erosion from a fertile soil is measured, the tons of total solids lost per acre do not give a picture of the seriousness of the loss. In such a case the concentration of plant nutrients in the runoff is determined. Since this concentration changes continuously and has no definite relationship to the rate of runoff, this determination must be made on an aliquot of the entire storm runoff, or many individual analyses of samples taken throughout the storm are needed to determine the losses of plant nutrients.

## *Measuring Accumulated Erosion*

A detailed survey of the remaining soil profile of an area discloses how much soil has been lost since the original soil was first disturbed by man. This can be done successfully only on soils that have definite horizons and where undisturbed profiles exist in the vicinity. In areas where cultivation has gone on for many centuries, as it has in many parts of China, it is difficult to find natural profiles for comparison.

This method can also be used to determine the amount of erosion for a given time period, if soil surveys are made at the beginning and at the end of the period. For such studies it is preferable not to depend on the change of the profile but to measure the lowering of the soil surface accurately with level and rod. Permanent elevations have to be established as bench marks, and seasonal changes in elevation, brought about by freezing, thawing, wetting, and drying, have to be taken into account.

A determination of the phosphate content of the soil at several depths can be used to study erosion and sedimentation. Surface soils are usually much higher in available phosphate than subsoils. If a survey indicates that the soil layer of relatively high phosphate content is thin or nonexistent, erosion has occurred. Where this layer is thicker than under natural conditions, sedimentation has added surface soil. This method of erosion study can be used where definite differences exist between the phosphate content of the surface and subsoils or where fertilization has enriched the plow layer only [36, 43].

In some exceptional cases the amount of erosion over a certain period can be estimated from the amount of sedimentation at the bottom of a watershed. Where a "soil-saving dam" has been installed or a rail fence or any other obstruction has been placed across the path of the runoff, most of the eroded soil is sedimented out, and its volume can be measured [25, 40]. Such sedimentation surveys are of no value, however, if the deposit is a considerable distance from the place where erosion has taken place. At any rate, never can all the eroded soil be found in such deposits, since some of it floats off suspended in water.

If the period of erosion is more than a few years, the amount of soil formation may be a sizable factor that has to be taken into consideration in making a survey-type study of accumulated erosion.

One basic difficulty in measuring erosion by any method is the fact that erosion is an intermittent process with soil removal and deposition alternating. The soil that reaches the measuring device or that is removed from a certain spot in the field may have moved downhill several times before.

## DAMAGE DONE BY WATER EROSION

The damage erosion causes is felt in various ways: it affects the land from which soil is washed; it damages the area downstream by floods and sediments; and it is a detriment to the economy because it lowers the income of the farmers and of the merchants and manufacturers who deal with the farmers.

In the following a brief summary is given of the damage done by water erosion:

*Disperses soil aggregates* through direct rainfall impact and removal of the binding substances. Detachment.

*Erodes internally* by washing soil particles into the cracks and pores of the soil body and making it less pervious to air and water.

*Decreases infiltration capacity and aeration capacity* through exposure of clay pan (*B* horizon) or through internal erosion.

*Increases runoff rates and amounts,* which results in more erosion hazards and in increased flood hazards.

*Removes soil,* especially topsoil, which usually contains more available plant nutrients and has a more desirable structure than the subsoil. Exposing tight subsoil generally impairs workability, root penetration, and water intake. Under most conditions subsoil cannot be improved to a degree that would make it as productive and as retentive for water as the original topsoil.

*Erodes differentially* by washing off the more valuable constituents of the soil, clay, organic matter, plant nutrients, seed, and young plants, leaving behind sand and stones.

*Cuts gullies,* thus depressing land-use capability.

*Decreases soil productivity* as a consequence of all previous items.

*Damages wildlife and fish* through decreased amount and quality of feed and muddying of water.

*Adds sediment,* filling up ponds, reservoirs, and streams and covering flood plains with increasingly worse soil materials.

## *Amounts of Erosion*

The actual amounts of erosion vary so greatly that it is possible only to give average amounts in broadest generalizations. According to Collier [15], soil loss from cropland in the American Middle West before World War II was at an annual rate of 3.9 tons per acre in the Plains states, 7.4 tons in the Lake states, and 15 tons in the corn belt. Losses from corn and other open-tilled crops are from 2 to 100 tons per acre per year; from small grains the losses are from 1 to 16 tons per acre per year; and from well-established meadows and well-managed pastures the losses are only a small fraction of a ton per acre per year. Losses from protected woodlands are usually negligible. Erosion losses of plant nutrients have been estimated by Lipman and Conibeare [49]. These losses and those from other causes, as well as the additions of plant nutrients, are shown in Table 3-12.

*Table 3-12. Balance Sheet of Additions and Losses of Plant Nutrients of Harvested Areas in the United States in 1930*
(In lb per acre, 367 million acres)

| | *N* | *P* | *K* | *Ca* | *Mg* | *S* |
|---|---|---|---|---|---|---|
| Additions* | 27.7 | 3.5 | 15.1 | 55.5 | 15.5 | 23.7 |
| Losses: harvested crops | 25.1 | 3.8 | 17.3 | 5.9 | 2.8 | 3.8 |
| Leaching | 23.0 | 0.0 | 37.6 | 151.2 | 34.8 | 41.8 |
| Erosion | 24.2 | 10.6 | 141.1 | 152.0 | 73.0 | 6.1 |

* Additions include fertilizer, liming materials, manure, bedding material, green manure, rainfall, irrigation, seeds.

The losses as stated by Lipman and Conibeare refer to the total losses of the various elements, not to the plant-available part only. This difference is particularly marked in the case of potassium. It has been pointed out before that an estimate of the amount of soil erosion can only be obtained for a given case if the movement of soil to be called erosion is first defined. Where we measure the entire splash due to the beating action of the raindrops, erosion may be as much as 150 tons per acre (1 inch) per storm. Where we measure the amount of soil that finds its way into a reservoir in the stream channel, it may be as little as 0.000082 in. per year [17, 27].

The results from the Lafayette, Indiana, watershed experiments are cited [32] as an example of erosion measured as the total soil loss passing from watersheds of 2 to 5 acres area. Soil losses totaled about 5 tons per acre (4 tons during the corn year) for a three-year period in corn, wheat, and meadow, where the corn and the wheat were planted in straight rows and small rates of fertilization were used. Where contouring and ample fertilization were used, the same rotation lost 2 tons per acre for the same three-year period. In some areas extremely severe soil losses occur in a short time. Rockie and McGrew [58] report that in one very intensive storm in the Palouse area of Washington 20,000 acres of summer fallow lost 275 tons per acre. In some areas they found 700 tons of soil per acre washed away, while in the main path of the water the entire plow depth was removed (about 1000 tons per acre). No accurate record of the amount of rainfall of this storm exists, but it is estimated to have exceeded 3 in. in a short period.

## SOIL EROSION BY WIND

### The Wind-erosion Cycle

The different phases of wind erosion from the first loosening of the soil to its deposition and consolidation can reoccur on the same field or for the same soil, so that these processes can be considered as a cycle. The phases are as follows [7]:

*Soil Loosening, Disintegration, and Denudation.* This is brought about by frost, wetting and drying, beating raindrops, and cultivation.

*Initiation of Movement.* The movement starts with the most erodible fractions. These are generally the particles between 0.1 and 0.5 millimeter in diameter.

*Transportation.* Transportation of soil by wind consists of three types of movement: surface creep, saltation, and suspension. These are discussed in detail later.

*Elutriation* (Sorting). The soil particles in transit are separated as a consequence of the effects of their size, density, and shape.

*Abrasion.* The process of soil particles hitting clods or stones and breaking them up into sizes that can be transported by the wind is called *abrasion.*

*Avalanching.* As soil particles fall down on the ground and cause others to start moving, more and more of the soil will be transported by wind the farther the wind blows across a field [8]. This increase in erosion is called *avalanching.*

*Deposition.* Corresponding to sedimentation in the case of erosion by water, wind-eroded soil is deposited in a variety of ways. The smallest and lightest particles of dust are carried the farthest, frequently far beyond the area where they were picked up. They only settle down when the wind subsides or when it rains. The larger particles that are moved in suspension (0.01 to 0.1 mm in diameter) are mostly deposited within a few miles of their origin. Under natural conditions these particles have formed the main loess deposits. These are not apt to be blown away, because they contain only a small proportion of saltation-size particles.

Particles from 0.1 to 0.5 mm in diameter that move by saltation are deposited wherever local conditions cause a reduction in wind intensity. This may be due to vegetation or to a slight unevenness of the land surface. The accumulation of these particles causes the formation

**Fig. 3-34.** Deposition of drifted soil. The drifted soil has come mostly from the field which is located across the highway (right). The ditch and the accumulation of tumbleweed in the fence have slowed down the wind and caused the soil to be deposited.

of hummocks or dunes. The soil grains that are moved by surface creep are carried a very short distance and remain in the general area where they originated.

*Soil Consolidation and Stabilization.* Once soil particles are deposited from the air, natural agencies take over to consolidate and stabilize the newly formed soil. Gravity, raindrop impact, the shrinking of wet soil when it dries, vegetation, and microflora and microfauna are mainly responsible for this.

## *The Forms of Soil Movement Caused by Wind*

**Saltation.** When strong winds sweep over unprotected land, some of the soil is picked up and carried away. The movement of soil is started by sand-sized particles rolling along the surface [6, 7]. The wind velocity at the very surface of the soil is practically zero, although at a small fraction of a millimeter above the ground, it may be quite considerable. This gives the soil particle much more impact at the top than at the bottom. The result is that the particle starts to

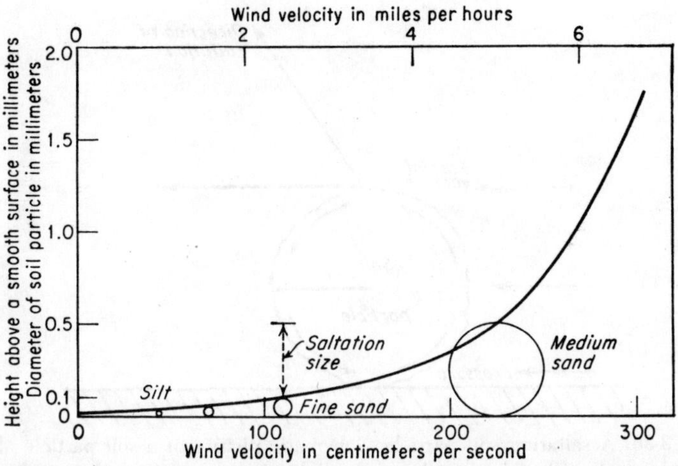


**Fig. 3-35.** Change in wind velocity with distance above the surface. Friction slows down the wind at the very surface, so that it cannot initiate the movement of particles of silt size. While soil grains larger than ½ mm diameter protrude enough to be driven by the wind, they spin too slowly for saltation to occur. Their weight increases as the cube of the diameter, while the increase in wind velocity beyond ½ mm diameter is very gradual. The velocities are given for a wind just strong enough to cause erosion.

spin at great speed, 200 to 1000 revolutions per second. As the particle is rolled along the surface of the ground, the top moves much faster than the wind, and the bottom part moves in opposite direction. As the air at the surface spins with the grain, a partial vacuum is created above the particle, and air is compressed below it. Both of these pressure changes tend to lift the particle. The grain jumps up almost vertically into the air. Its horizontal momentum causes it to rise at an angle between 75 and 90°. It climbs to 6 to 12 in. in height, sometimes to 2 or even 3 ft. As it is lifted up, the spin slows down, and the particle enters layers where the wind is substantially faster. Having thus lost its upward impulse, it is carried with the wind in slightly descending direction. The angle with which it reaches the ground is between 6 and 12°.

This form of wind erosion by short jumps is called *saltation.* Only a rather narrow range of soil-particle sizes of the total soil mass is suitable to be caught by the wind and to develop enough difference in air pressure above and below the particle. They must be small

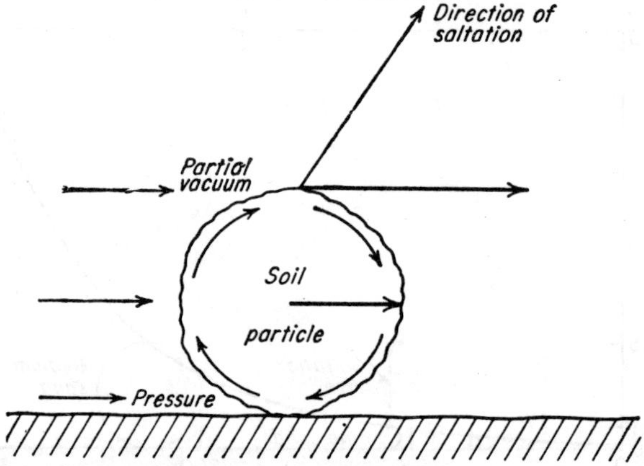


**Fig. 3-36.** A saltating soil particle. Schematic picture of a soil particle about to be lifted into the air. The lengths of the light arrows represent the wind velocity, those of the heavy arrows the forward motion of the surface of the rapidly rolling particle. Since its motion at the top is greater than that of the wind, a partial vacuum is created above the particle. The slow motion at the bottom of the particle causes air to be compressed by the wind. Reduced pressure above and increased pressure below result in the lifting of the particles.

enough that the suction developed by the spinning is sufficient to lift them up. Most soil particles moved by saltation are between 0.1 and 0.5 mm in diameter. This is in the range of fine sand.

**Suspension Movement.** When the particles that move by saltation drop down, they stir up the soil at the spot they hit. The smaller soil particles are thus kicked into the air, and because of their relatively large surface compared to their weight, they stay in suspension and are carried long distances by the wind [6]. Such dust sometimes travels several thousand miles before it again settles down to earth or into the ocean. It is composed of particles smaller than 0.1 mm in diameter.

It is an interesting phenomenon that a layer of such dust is not raised into the air directly by the wind. The particles are so small that they do not protrude enough from the general level of the ground for the wind to move them. It takes the impact of a descending particle to cause dust to blow in an ordinary wind. This is the reason that loess deposits have developed, instead of being driven away by the next wind. Loess is made up largely of silt-size particles of 0.05 to 0.002 mm. The loess was originally deposited during windstorms which separated it from both coarser and finer particles.

**Surface Creep.** Soil particles larger than 0.5 mm in diameter are too heavy to be lifted by wind. They are pushed along the surface of the ground, rolling, but not jumping or flying [6, 9]. This movement is called *surface creep*. The kinetic energy of strong winds is sufficient to cause this movement. When the wind is less severe, surface creep is initiated by saltation. The descending soil particles hit the larger grains and impart their energy to them, similar to the action of two billiard balls. Soil grains larger than 3 mm diameter are too large to be moved by ordinary winds.

While there are no hard and fast boundaries between the sizes of soil particles that move in the three forms of wind erosion, because of the differences in wind velocity and turbulence and because of the variations in shape and density of soil particles themselves, Table 3-13 gives a general idea of these relationships.

*Table 3-13. Approximate Sizes of Soil Particles Moved by Wind*

| Form of wind erosion | Diameter of soil particles |
|---|---|
| Suspension flow | Smaller than 0.1 mm |
| Saltation | 0.1–0.5 mm |
| Surface creep | 0.5–3.0 mm |

Particles larger than 3 mm diameter are only moved by tornadoes.

## *Factors Affecting Soil Erosion by Wind*

**Climate Air Velocity and Turbulence.** Wind velocity at the ground surface is the most important factor. There is a minimum wind velocity to start erosion for any field condition. If no soil particles are in the air, a higher velocity is required to initiate the movement than when saltation particles from a neighboring area are falling down on the ground and adding their kinetic energy to that of the wind to start the soil moving. This latter is called the *minimal-impact* *threshold velocity,* while the wind velocity required to start erosion without assistance of outside particles is called the *minimal-fluid threshold velocity*. Depending upon the given situation this is somewhere between 8 and 13 miles per hour, measured 6 in. above the ground [6].

Turbulence adds a vertical component to the wind and makes it more erosive. The ability of the wind to move and pick up soil particles varies at least as the second or possibly as the third power of the velocity.

A wind-velocity scale is given in Table 3-14. It should be kept in mind, however, that wind velocities are commonly measured several feet above the ground, while it is velocity against the soil particles that concerns us in wind-erosion studies. The velocity of the wind hitting the soil particles that project above the average ground level is much lower than at higher levels.

In 1806 Admiral Beaufort established a scale of wind velocity according to its effect on various objects. Wind direction and velocity vary with the distance above the ground, the greatest variation being near the ground. The wind velocity as given in the Beaufort scale is for wind at the standard anemometer elevation of 33 ft (10 m) above the ground.

*Temperature.* Temperature, barometric pressure, and relative humidity affect the specific gravity of air and consequently the energy with which wind hits the soil. These effects are probably only small. Temperature and relative humidity, through their effect on evaporation, are of great importance in soil erosion.

*Precipitation.* The amount and the distribution of rainfall have a strong indirect effect on wind erosion by providing the ground with the water necessary for a protective plant cover and by keeping the soil moist and cohesive.

**Soil.** The important factors affecting the erodibility of soil by wind are its texture, structure, cohesiveness, bulk density, and moisture content at the surface.

Sands erode easily because they contain a great proportion of particles of saltation size, but little binding material. Soils high in clay and silt are rather stable because they are coherent and easily form a protective crust at the surface. Mucks erode readily because of their low density. Soils that contain aggregates of erodible size erode

*Table 3-14. Wind Velocities, Beaufort Scale*

| Beaufort number | Descriptive word | Velocity, mph | Specifications for estimating velocities | Wind-erosion hazard |
|---|---|---|---|---|
| 0 | Calm | Less than 1 | Smoke rises vertically | |
| 1 | | 1–3 | Direction of wind shown by smoke but not by wind vanes | None |
| 2 | Light | 4–7 | Wind felt on face; leaves rustle; ordinary vane moved by wind | |
| 3 | Gentle | 8–12 | Leaves and small twigs in constant motion; wind extends light flag | Begins in muck |
| 4 | Moderate | 13–18 | Raises dust and loose paper; small branches are moved | Slight |
| 5 | Fresh | 19–24 | Small trees in leaf begin to sway; crested wavelets form on inland water | |
| 6 | | 25–31 | Large branches in motion; whistling heard in telegraph wires; umbrellas used with difficulty | Considerable |
| 7 | Strong | 32–38 | Whole trees in motion; inconvenience felt in walking against the wind | |
| 8 | | 39–46 | Breaks twigs off trees; generally impedes progress | |
| 9 | Gale | 47–54 | Slight structural damage occurs (chimney pots and slate removed) | Severe |
| 10 | | 55–63 | Trees uprooted; considerable structural damage occurs | |
| 11 | Whole gale | 64–75 | Rarely experienced; accompanied by widespread damage | |
| 12 | Hurricane | Above 75 | | |

Partially taken from B. C. Haynes, Meteorology for pilots, *Civil Aeronaut. Bull.* 25, January, 1943.

Conversion factors: 1 mph = 44.7 cm per sec.
100 cm per sec = 2.24 mph.

easily because aggregates have a low bulk density (volume weight) and therefore are lifted by winds of fairly low velocity. Coarsely granulated soils erode mostly in saltation, finely pulverized soils in saltation and suspension. Soils containing larger aggregates and those with a surface crust are rather resistant to erosion, as are soils composed exclusively of fine dust, so long as no particles in saltation impinge upon them.

Under practical conditions, however, the finer the soil is, the more it blows, because of the smoothness of the ground and because there are generally enough particles of 0.1 to 0.5 mm diameter to start saltation and to raise dust.

Wind erosion can only happen when the soil surface is dry or very slightly moist, because surface tension holds the soil particles together if they are moist or wet. Soils that have a tendency to retain moisture and to conduct it to the surface are therefore fairly resistant to drifting. A dry wind, however, can dry out the soil surface very fast. In semiarid regions intense wind erosion has been observed an hour after a heavy rainfall.

**Ground Conditions and Land Use.** *Topography.* Level land is generally more liable to wind erosion than rolling land because the wind finds less resistance. Nevertheless, some of the greatest erosion hazards exist on knolls, ridges, and in the lee of pockets, because the wind presses against such areas instead of flowing parallel to the surface.

*Microtopography.* Small depressions catch saltating particles, thus protecting soil from wind erosion [69]. Hummocks formed by deposits of saltating particles erode particularly easily, since they stick out from the general ground level and are therefore exposed to stronger wind.

*Plants.* Plants slow down the wind near the surface, and their shade helps to keep the soil surface moist. Therefore soil drifting in a dense stand of plants hardly ever occurs. The longer the plants cover the soil during the year, the less it erodes.

*Crop Residues.* The threshold wind-erosion velocity of crop residues, even if they lie loose on the ground, is fairly high [6, 21]. Therefore they help to slow down winds of medium velocity at the ground level. Strong winds blow away such residues but will not move stubble or other material that is anchored to the soil. Stubble is one of the most efficient traps for saltating soil particles.

*Mechanical Obstructions.* Depending upon their shape, location, and permeability to air, mechanical obstructions may decrease wind velocity near the ground, or they may concentrate it there.

*Tillage.* Soil is liable to be eroded by strong winds when it is bare and dry and when it contains many particles of erodible size at the surface. Tillage brings about such conditions and therefore has made soil vulnerable to wind erosion by exposing bare soil, by drying it out, and by oxidizing the humus. The smoother and finer the ground, the greater is the erosion hazard. Therefore the ground is sometimes roughed up and made cloddy with a lister or chisel-point cultivator as an emergency measure against wind erosion.

*Pasturing.* Where grazing is controlled so that a fair cover of grasses remains, erosion is of little consequence. Overpasturing, on the other hand, is a frequent cause of erosion in semiarid regions. Frequently, water and wind erosion combine to denude such areas of their topsoil and lower their fertility, so that grasses disappear and a "short-grass prairie" changes to desert.

**Season.** The wind-erosion hazard varies with the season. The occurrences of strong drying winds, of a rainy season, of freezing weather, and of the cycle of agricultural operations are some of the reasons for this variation [11]. Most of these factors vary from area to area and are rather obvious in their effects, so that they need not be discussed here. The effect of freezing on the erodibility of soil is not so readily recognized. Where moist soils freeze, the coarse, water-stable aggregates tend to break down, and at the same time, the finest fractions consolidate to an intermediate size, especially between 0.05 and 0.4 mm diameter [12]. Therefore such soils can readily be moved by saltation and are much more erodible than they were before freezing. This effect is particularly pronounced in heavy soils and only of minor importance in sandy soils. Where the climate is such that the soils are dry during the winter, very little effect of freezing temperatures on erodibility has been found.

## *Man-induced Wind Erosion*

Wind erosion occurs only where the land is not protected by sufficient vegetation. Under undisturbed natural conditions this is the case in arid and semiarid regions, on the shores of oceans and lakes, and in rocky, mountainous areas. By removing or reducing the vegetative cover man has caused wind erosion to spread to many more areas.

Wind erosion occurs wherever agriculture is practiced, but in many cases such erosion is inconsequential. It is serious where dry soil of the critical size is exposed to strong winds. Wind erosion is serious in arid and semiarid regions and on muck and sand. Since we are concerned only with man-induced, accelerated wind erosion, we can exclude the arid areas, because only irrigation agriculture is practiced there.

The most serious wind-erosion hazard exists on cultivated land in semiarid and subhumid regions. The classic example in the United

**Fig. 3-37.** Wind-erosion damage. High-velocity wind has severely damaged this South Dakota field which was prepared for wheat seeding. Much of the fertile topsoil is piled along the fence. (*USDA*)

States is the Dust Bowl in western Kansas, Oklahoma, and Texas [34]. Here, land use has been adopted that is not suited to the climatic conditions. Many years of continuous cropping to wheat and exposing the soil through the practice of summer fallow have created a severe wind-erosion hazard. Favorable weather conditions and more careful farming have minimized soil drifting in that area. But wind erosion is still common in the Great Plains of the North American continent from the Canadian prairie provinces to Mexico. In other areas, for instance, the Navajo Indian reservation in northwestern New Mexico and the Near East from the Mediterranean to Pakistan, overpasturing has robbed the soil of its protective cover.

No serious, large-scale wind erosion occurs in Europe, because the climate is moist. North Africa, Arabia, and much of the Near East suffer from wind erosion. This is owing to the centuries of cultivation and the hot and semiarid climate of that region. Frequently, it is difficult to say whether wind erosion is man-induced or not. It is

**Fig. 3-38.** Wind-erosion damage. Wind erosion such as this causes serious property damage and inconvenience. (*USDA*)

probable that the erosion in the deserts of North Africa and Arabia is of strictly natural origin, although the geographic extent of these deserts has shifted because of man's activities. Other areas of serious wind erosion are in South Africa, Australia, and southeastern Russia.

## *Damage Done by Wind Erosion*

Wind erosion causes damage in a variety of ways. The finest and most valuable parts of the soil, clay, silt, and organic matter are moved the farthest, usually away from the farm where they were picked up. The soil that remains becomes increasingly more coarse textured and less fertile. The accumulation of saltation-size sand in hummocks makes farming more difficult and helps to start erosion when the next strong wind hits.

Soil drifting sometimes cuts off young seedlings, and the wind may actually blow seeds away or cover them up so deeply that they cannot develop [73]. In the same way, weed seeds may be spread from field to field.

Roads, ditches, railroads, and fences are covered by soil that requires expensive rehabilitation. Soil gets into engines and other moving parts of automobiles, tractors, and farm implements, thus wearing them out prematurely. Dust and dirt invade houses and even get into food, making life unpleasant. The most direct and serious damage caused by dust storms is the impairment of human and animal health. Dust pneumonia and inflamed eyes are frequently encountered after severe wind erosion.

## SUBSIDENCE OF ORGANIC SOILS

A special form of erosion is the subsidence of organic soils. It is debatable whether this should be considered erosion, since by and large, no mechanical removal of soil occurs.

Where conditions for plant growth are favorable and decomposition of organic matter is impeded by lack of oxygen or low temperatures, peat and muck soils develop. In temperate and tropical climates organic soils develop only where water excludes oxygen; otherwise, conditions are favorable for plant growth. This means that a steady water level is a prime requirement for the formation of mucks and peats outside the arctic regions. Whenever geologic erosion has caused a continuously sloping drainage pattern, water levels fluctuate, permitting oxygen to enter the soil and to decompose plant residues. This is the reason that organic soils occur particularly in areas that have been covered by glaciers in recent geologic times and contain undrained "kettle holes." The few areas of organic soils in the tropics and the subtropics occur where some specific geologic condition exists that maintains water levels at the ground surface, as is the case in Florida and in the Dominican Republic.

In cold climates the rate of decomposition is so slow that plant residues cover slopes and ridges as well as depressions with peat deposits. During a short time in the summer the air temperatures are sufficiently high to cause rapid plant growth; the soil, however, stays cool enough to retain much of the plant material.

Subsidence is a serious problem whenever organic soils are drained, as they have to be for the production of most agricultural crops. This lowering of the ground-water table brings about several conditions that contribute to a lowering of the land surface. Drying out organic soils causes them to shrink. Once they have become as dry as the air around them, they will not expand to the full original volume, even after long submersion. In other words, this shrinkage is only partially reversible.

More important than shrinkage by dehydration is the oxidation of the organic substance with the end products of carbon dioxide, water, and minerals. This oxidation is responsible for the greatest part of the subsidence of organic soils. The rate increases with the average annual temperature, with the depth of drainage, and with the content of organic matter in the soil. It is also greatly affected by the botanical and chemical composition of the organic soil. The annual amount of shrinkage of muck soils in Indiana has been found to be around 0.6 inch where the water table during the growing season has been lowered to 24 in. below the surface [35]. In Florida, for the same depth of water table, the subsidence is 1.4 in. per year [63]. In fact the part of the Florida Everglades that was covered with about 12 ft of peat in 1914 when drainage started has already lost 6 ft of peat, and it is anticipated that by A.D. 2000 all the organic soil will have disappeared if the present form of agriculture continues. It has been suggested [26] that the area be kept under water and used for rice production. No other method of conserving organic soils has been found. This destruction can be slowed down by maintaining water levels near the surface, but if the soils are drained enough for the growth of upland crops, shrinkage continues.

In addition to dehydration and oxidation, fire and wind erosion can cause the loss of organic soils.

Maintenance of organic soils or of the organic covering of mineral soils is not beneficial in some cases of subarctic agriculture. Plant residues contain much of the nutrient elements and do not release them until they are oxidized. Under such exceptional circumstances careful burning of the surface layers is advisable.

## REFERENCES


1. Baver, L. D.: Ewald Wollny—A pioneer in soil and water conservation research, *Soil Sci. Soc. Am. Proc.*, **3**:330–333, 1938.
2. Borst, H. L., A. G. McCall, and F. G. Bell: Investigations in erosion control and the reclamation of eroded land at the northwestern Appalachian Conservation Experiment Station, Zanesville, Ohio, 1934–1942, *USDA Soil Conservation Service Tech. Bull.* 888, 1945.
3. Borst, H. L., and R. Woodburn: The effect of mulching and methods of cultivation on runoff and erosion from Muskingum silt loam, *Agr. Eng.*, **23**:19–22, 1942.
4. Bouyoucos, G. J.: The clay ratio as a criterium of susceptibility of soils to erosion, *J. Am. Soc. Agron.*, **27**:738–741, 1935.
5. Brown, C. B.: Erosion control on watershed lands, *J. Am. Watersheds Assoc.*, **38**:1127–1137, 1946.
6. Chepil, W. S.: Dynamics of wind erosion: I. Nature of movement of soil by wind, *Soil Sci.*, **60**:305–320, 1945.
7. Chepil, W. S.: Dynamics of wind erosion: II. Initiation of soil movement, *Soil Sci.* **60**:397–411, 1945.
8. Chepil, W. S.: Dynamics of wind erosion: V. Cumulative intensity of soil drifting across eroding fields, *Soil Sci.*, **61**:257–263, 1946.
9. Chepil, W. S.: Dynamics of wind erosion: VI. Sorting of soil material by the wind, *Soil Sci.*, **61**:331–441, 1946.
10. Chepil, W. S.: Factors that influence clod structure and erodibility of soil by wind: Organic matter at various stages of decomposition, *Soil Sci.*, **80**:413–420, 1955.
11. Chepil, W. S.: Field structure of cultivated soils with special reference to erodibility by wind, *Soil Sci. Soc. Am. Proc.*, **17**:185–191, 1953.
12. Chepil, W. S.: Seasonal fluctuations in soil structure and erodibility of soil by wind, *Soil Sci. Soc. Am. Proc.*, **18**:13–16, 1954.
13. Clark, O. R.: Interception of rainfall by prairie grasses, weeds, and certain crop plants, *Ecol. Monograph*, **10**:243–277, 1940.
14. Clarke, F. W.: The composition of the river and lake waters of the United States, *U.S. Geol. Survey Profess. Paper* 135, 1924.
15. Collier, G. W.: Soil conservation during the war, *USDA Soil Conservation Service War Records Monograph* 2, March, 1946.
16. Duley, F. L., and O. E. Hays: The effect of the degree of slope on runoff and soil erosion, *J. Agr. Research*, **45**:349–360, 1932.
17. Eakin, H. M.: Silting of reservoirs, *USDA Tech. Bull.* 524, 1936.
18. Ellison, W. D.: Soil detachment and transportation, *Soil Conservation*, **11**:179, 1946.
19. Ellison, W. D.: Soil erosion, *Soil Sci. Soc. Am. Proc.*, **12**:479–484, 1947.
20. Ellison, W. D.: Protecting the land against the raindrop's blast, *Sci. Monthly* 68, April, 1949.

21. Englehorn, C. L., A. W. Zingg, and N. P. Woodruff: The effect of plant residue cover and clod structure on soil losses by wind, *Soil Sci. Soc. Am. Proc.*, **16**:29–33, 1952.
22. Eser, Carl: Untersuchungen ueber den Einfluss der physikalischen und chemischen Eigenschaften des Bodens auf dessen Verdunstungsvermoegen, *Fortschr. Gebiete Agrikulturphysik*, **7**:46–99, 1884.
23. Garstka, W. U.: Design of the automatic recording in-place lysimeters near Coshocton, Ohio, *Soil Sci. Soc. Am. Proc.*, **2**:555–556, 1937.
24. Geib, H. V.: A new type of installation for measuring soil and water losses from control plots, *J. Am. Soc. Agron.*, **25**:429–440, 1933.
25. Gerdel, R. W.: Soil losses from cultivated strips in strip cropped fields in the Ohio Valley Region, *USDA Circ.* 588, 1940.
26. Green, V. E.: Rice—soil conserving or soil depleting?, *Soil Sci. Soc. Am. Proc.*, **17**:283–284, 1953.
27. Happ, S. C., and G. C. Dobson: Some principles of accelerated stream and valley sedimentation, *USDA Tech. Bull.* 695, 1940.
28. Harrold, L. L.: Hydrologic design of farm ponds and rates of runoff for design of conservation structures in the North Appalachian Region, *USDA SCS-TP*, October, 1946.
29. Harrold, L. L.: Report of committee on infiltration, 1950–1951, *Trans. Am. Geophys. Union*, **32**:919–922, 1951.
30. Harrold, L. L., and F. R. Dreibelbis: Agricultural hydrology as evaluated by monolith lysimeters, *USDA Tech. Bull.* 1050, 1951.
31. Hays, O. E., A. G. McCall, and F. G. Bell: Investigations in erosion control and the reclamation of eroded land at the upper Mississippi Valley conservation station near LaCrosse, Wisconsin, 1933–1943, *USDA Tech Bull.* 973, 1949.
32. Hickok, R. B., I. D. Mayer, and H. Kohnke: Some runoff control and moisture conservation possibilities, *Agr. Eng.*, **29**:257–261, 1948.
33. Jenny, H.: "Factors of Soil Formation," McGraw-Hill Book Company, Inc., 1941.
34. Joel, A. H.: Soil conservation reconnaissance survey of the southern Great Plains wind-erosion area, *USDA Tech. Bull.* 556, 1937.
35. Jongedyke, H. A., et al.: Methods and effects of maintaining different water table levels in muck soil, *USDA and Purdue Univ. Agr. Expt. Sta.*, 1951. (Processed.)
36. Jung, L.: Untersuchungen ueber den Einfluss der Bodenerosion auf die Ertraege in haengigem Gelaende, *Bundesministerium fuer Ernaehr., Landwirtsch und Forsten, Schriftenreihe fuer Flurbereinigung,* 1956.
37. Jung, L.: Anlage zur Messung von Abfluss und Abtrag auf landwirtschaftlichen Nutzflaechen. *Schriftenreihe des Kuratoriums fuer Kulturbauwesen,* **5**:19–24, 1956.
38. Kohnke, H.: Runoff chemistry: An undeveloped branch of soil science. *Soil Sci. Soc. Am. Proc.*, **6**:492–500, 1941.

39. Kohnke, H., and R. B. Hickok: An automatic aliquot runoff sampler, *Soil Sci. Soc. Am. Proc.*, **8**:444–447, 1944.

40. Kohnke, H., and F. R. Dreibelbis: Methods of measuring soil erosion, *Soil Research*, **4**:232–241, 1939.

41. Kohnke, H., F. R. Dreibelbis, and J. M. Davidson: A survey and discussion of lysimeters and a bibliography on their construction and performance, *USDA Misc. Publ.* 372, 1940.

42. Krumbein, W. C., and L. L. Sloss: "Stratigraphy and Sedimentation," W. H. Freeman Company, San Francisco, Calif., 1951.

43. Kuron, H.: Veraenderungen der Ackerboeden unter dem Einfluss der Bodenerosion, *Z. Pflanzenernaehr. Dueng. Bodenk.*, **41**:245–258, 1948.

44. Kuron, H., L. Jung, und H. Schreiber: Messungen von oberflaechlichem Abfluss und Bodenabtrag auf verschiedenen Boeden Deutschlands, *Schriftenreihe des Kuratoriums fuer Kulturbauwesen,* **5**:88, 1956.

45. Lamar, W. L., and C. R. Collier: Suspended sediment characteristics of Ohio streams, *J. Soil and Water Conservation,* **11**:233–238, 1956.

46. Laws, J. O.: Measurements of the fall velocities of water drops and raindrops, *Trans. Am. Geophys. Union,* **22**:709–721, 1941.

47. Laws, J. O., and D. A. Parsons: The relation of raindrop size to intensity, *Trans. Am. Geoph. Union,* **24**:452–460, 1944.

48. Lillard, J. H., H. T. Rogers, and J. Elson: Effect of slope, character of soil, rainfall, and cropping treatments on erosion losses from Dunmore silt loam, *Virginia Agr. Expt. Sta. Tech. Bull.* 72, 1941.

49. Lipman, J. G., and A. B. Conibeare: Perliminary note on the inventory and balance sheet of plant nutrients in the United States, *New Jersey Agr. Expt. Sta. Bull.* 607, 1936.

50. Middleton, H. E.: Properties of soils which influence soil erosion, *USDA Tech. Bull.* 178, 1930.

51. Middleton, H. E., C. S. Slater, and H. G. Byers: Physical and chemical characteristics of the soil from the erosion experiment stations—first report, *USDA Tech. Bull.* 316, 1932.

52. Osborn, B.: Field measurements of soil splash, *J. Soil and Water Conservation,* **8**:255–260, 1953.

53. Parshall, R. L.: The Parshall measuring flume, *Colo. Expt. Sta. Bull.* 423, 1936.

54. Parsons, D. A.: Coshocton-type runoff samplers, *USDA Agr. Research Service,* 1955, pp. 41–42.

55. Peterson, J. B.: Calcium linkage, a mechanism in soil granulation, *Soil Sci. Soc. Am. Proc.,* **12**:29–31, 1947.

56. Pittman, D. D., and H. Kohnke: An automatic self-recording infiltrometer, *Soil Sci.,* **53**:429–434, 1942.

57. *Purdue Univ. Agron. Handbook,* 1956.

58. Rockie, W. A., and P. C. McGrew: Erosive effects of heavy summer rains in southeastern Washington, *Wash. Agr. Expt. Sta. Bull.* 271, 1932.

**59.** Smith, D. D., D. M. White, A. W. Zingg, A. G. McCall, and F. G. Bell: Investigations in erosion control and reclamation of eroded Shelby and related soils at the conservation experiment station at Bethany, Missouri, 1930–1942, *USDA Tech. Bull.* 883, 1945.

**60.** Smith, F. B.: The effect of organic matter on aggregation, permeability and runoff, *J. Soil and Water Conservation,* **10**:76–80, 1955.

**61.** Soil and water conservation research branch agricultural research service and soil conservation service, minutes of joint conference on slope-practice held at Purdue University, July 20–23, 1956.

**62.** Stall, J. B., A. A. Klingebiel, S. W. Melsted, and E. L. Sauer: The silting of Lake Calhoun. Illinois State Water Survey Division, *Rept. of Investigation No.* 15, 1952.

**63.** Stephens, J. C.: Subsidence of organic soil in Florida Everglades, *Soil Sci. Soc. Am. Proc.,* **20**:77–80, 1956.

**64.** Thomas, H. E.: First fourteen years of Lake Mead, *U.S. Geol. Survey Circ.* 346, 1954.

**65.** Thorp, James: Geography of the soils of China. *Natl. Geol. Survey China,* Nanking, 1936.

**66.** U.S. Bureau of Reclamation: Study of variability of sand deposits, *U.S. Waterways Exp. Sta. Misc. Paper,* 1955, pp. 3–12.

**67.** U.S. Department of Commerce: Rainfall intensity–duration–frequency curves, *Weather Bur. Tech. Paper* 25, 1955.

**68.** U.S. Weather Bureau: Maximum 24-hour precipitation in the United States, *Tech. Paper* 16, 1952.

**69.** Van Doren, C. E.: The effect of cloddiness of soils on their susceptibility of wind erosion, *J. Am. Soc. Agron.,* **36**:859–864, 1944.

**70.** Wilm, H. G.: The application of measurement of artificial rainfall on types FA and F infiltrometers, *Trans. Am. Geophys. Union,* **24**:480–487, 1943.

**71.** Wischmeier, W. H., and D. D. Smith: Rainfall energy and its relationship to soil loss, *Trans. Am. Geophys. Union,* **39**:285–291, 1958.

**72.** Wischmeier, W. H.: A rainfall erosion index for a universal soil-loss equation, to be published in *Soil Sci. Soc. Am. Proc.*

**73.** Woodruff, W. P.: Wind blown soil abrasive injuries to winter wheat plants, *Agron. J.,* **48**:499–504, 1956.

**74.** Yarnell, D. L.: Rainfall intensity–frequency data, *USDA Misc. Publ.* 204, 1935.

# 4

# *Aims and Principles of Soil Conservation*

## PURPOSES OF SOIL CONSERVATION

It is the aim of soil conservation to obtain the greatest possible permanent benefit from the land. This means using the land and at the same time maintaining or improving its productive capacity. The products may be agricultural or horticultural crops, pasture, forest, wildlife, or water, or they may be purely aesthetic. Frequently they are a combination of several of these. The specific aims of soil conservation may be:

1. To keep erosion down to a rate where natural soil development together with cultural practices can keep a balance with the productivity loss brought about by erosion
2. To bring the concentrations of plant nutrients in the soil to a satisfactory level, to protect them from unnecessary losses, and to replace those lost by any cause
3. To maintain organic matter by proper adjustment of the oxidation rate and to replenish these organic substances
4. To retain or to improve the tilth of the soil
5. To make the best use of the available water

These five aims are so closely interrelated that it is difficult to discuss one without discussing the others. Both soil erosion and soil conservation are cumulative processes; the more the soil has eroded, the more it has the tendency to erode. By the same token the more conservation measures are employed, the safer from erosion will the soil become over the years.

Generally, it is difficult, if not impossible, to bring a severely eroded soil to the productivity it could have reached if it had not been eroded, regardless of the cultural improvements employed. It is a better policy to conserve a soil than to attempt its reclamation.

There is no short-cut method, no panacea to accomplish soil conservation. Each case has to be analysed, and usually only a combination of several measures will bring about the desired result.

It should be made entirely clear that the aim of conservation is wise use, not preservation. It is not the intent to make the soil a museum piece and to maintain it in the condition in which it was first discovered. The soil, as one of our most important natural resources, has to provide mankind with many of the necessities of life.

## *Main Approaches to Soil Conservation*

Whatever the techniques employed, four different approaches are used to conserve the soil:

1. Conditioning the soil to make it resistant to detachment and transportation and more absorptive for surface water
2. Covering the soil to protect it from rainfall impact and wind
3. Slowing down runoff and wind
4. Providing safe ways for the disposing of unavoidable surface runoff

To condition the soil so that it is resistant to erosion and as productive as possible, its structure has to be both favorable to plant growth and resistant to deterioration, and its moisture content has to be intermediate, so that it allows for ready infiltration of rain water and there are both water and air at the disposal of the plant roots. Nutrients should be available at all times at the rates the plants need them for best growth.

Covering the soil by plants or plant residues protects the soil from the impact of rain, reduces wind velocity at the ground level, and helps to condition the soil.

Slowing down runoff and wind reduces detachment and transportation of soil particles. This retarding is achieved by increasing infiltration capacity and surface storage and by creating obstructions to runoff and wind. These may be plants and plant residues, or they may be purely mechanical. Avoiding concentrations of runoff and of wind are also necessary to slow them down.

## PERMISSIBLE EROSION

As it is practically impossible to farm without any loss of soil by erosion, it is necessary to establish limits of permissible erosion for the many cases of soils and climates. These limits have to be selected in such a way that full productivity can be maintained forever and that no damage results to neighboring areas. To arrive at these limits several items have to be considered.

Soil is continuously formed by nature. This rate is speeded up by man. Erosion should not exceed the rate of anticipated soil formation. An adequate depth of the soil must be maintained, so that enough water can be stored and the roots find sufficient nutrients, water, and mechanical support. This means that a soil with underlying material that readily changes to soil can be allowed to have more erosion than one with a hardpan or resistant rock near the surface. Plant nutrients and organic matter should not be removed at a rate greater than they can be replaced.

Erosion should be held below the rate at which gully formation is likely to occur. The sediment from an area that erodes at a rate higher than 5 tons per acre per year fills up large reservoirs at an excessive rate. In the case of small reservoirs the permissible rate of erosion is even smaller. In a similar way wind erosion should be kept low enough so that its deposits do not fill ditches or cover crops, equipment, and roads. It should also not cause blowout holes or hummocks.

To determine the limits of permissible erosion with any degree of accuracy is very difficult, because several of the items that should be used in arriving at these limits cannot, or have not, been determined quantitatively. The figures that are used at present are based largely on the maintenance of productivity of the runoff plots on the soil and water conservation research stations of the U.S. Department of Agriculture. They range from 1 to 5 tons per acre per year, depending on the permeability and the depth of the soil profile and to what extent it is already eroded. This is in line with the observation that the rate of soil formation from glacial-till parent material near the surface is estimated to be around 3 tons per acre per year in the American Middle West. No limits of permissible wind erosion have been established so far.

### *Practical Use of Limits of Permissible Soil Loss*

Once the limit of permissible erosion has been determined for a certain soil, the land use has to be planned in such a way that the rate of erosion does not exceed this limit. The rate of erosion $A$ in a given climate is affected by the land use or crops grown $C$; the level of management $M$, for instance, liming, fertilization, drainage, and tillage; the steepness of the slope $S$; the length of the slope $L$; specific erosion-control practices used $P$; and the erodibility of the soil $E$, which itself is affected by the amount of previous erosion.

The "erosion equation" that relates all these factors to the rate of erosion is

$$A = CMSLPE$$

where $A$ = the estimated soil loss, tons per acre per year
$C$ = the crop-rotation factor, tons per acre per year
$M$ = the management factor
$S$ = the slope-gradient factor
$L$ = the slope-length factor
$P$ = the conservation-practice factor
$E$ = the soil-erodibility factor (this includes a factor based on previous erosion)

Numerical values for these factors have been developed for the main agricultural regions of the United States. In each case a specific condition is established as a base. By relating the situation as it exists in a given field to this base condition and fitting all the factors into the erosion equation, an estimate of the anticipated annual soil loss can be obtained. This is a very useful technique to determine whether or not a proposed combination of land use and management will result in a rate of erosion that will be within permissible limits. If the anticipated erosion is too large, an adjustment has to be made that will decrease one or several of the factors and consequently lower the anticipated erosion.

In the following paragraphs the details of the erosion-equation technique are illustrated by the use of the factors as recommended for the Middle Western United States.*

* These data are based on a mimeographed report of a workshop on slope-erosion by representatives of the Soil Conservation Service, soil and water con-

*Table 4-1. Soil-loss Factors for Crops under Various Cropping Systems, Based on Loss from Continuous Row Crops**

| Crop | Cropping factor |
|---|---|
| Continuous row crop | 1.00 |
| Row crop after one year meadow | 0.40 |
| Row crop after two or more years meadow | 0.35 |
| Row crop after small grain | 0.90 |
| | |
| Row crop after row crop after small grain | 1.00 |
| Row crop after row crop (or grain) after one year meadow | 0.80 |
| Row crop after row crop (or grain) after two or more years meadow | 0.70 |
| Third year row crop after one year meadow | 0.95 |
| Fourth or more year row crop after one year meadow | 1.00 |
| | |
| Spring grain after one or two years row crop (or grain) after catch crop | 0.35 |
| Spring grain after row crop (or grain) after one year meadow | 0.30 |
| Spring grain after row crop (or grain) after two or more years meadow | 0.25 |
| Spring grain after second-year row crop (or grain) after one year meadow | 0.35 |
| Spring grain after second-year row crop (or grain) after two or more years meadow | 0.30 |
| Spring grain after three or more years row crop (or grain) | 0.40 |
| Spring grain after one year meadow | 0.15 |
| Spring grain after two or more years meadow | 0.10 |
| | |
| First-year meadow | 0.010 |
| Second-year and succeeding meadow | 0.005 |
| Lespedeza hay | 0.040 |

* Fertility treatments sufficient to produce good yields are assumed in these factors.

As a basis for calculations the average annual erosion from a three-year rotation of row-crop—spring-grain—meadow is set as 8 tons per acre, assuming all other factors are unity. This is a somewhat unfortunate choice of a base land use, since continuous row crop is selected to have a cropping factor of unity.

*C, the Crop-rotation Factor.* The soil loss from continuous row crop (generally corn) has been given the value of 1.00. If loss from continuous row crop is 20 tons per acre and the loss from oats in a rotation is 5 tons per acre, then the *C* factor for oats is 0.25. Soil-loss factors for most of the important crop combinations in the Middle West are given in Table 4-1.

servation branch of the Agricultural Research Service, State Experiment Stations, and Extension Service at Purdue University, Lafayette, Ind., July, 1956.

*M, the Management Factor.* Conventional management without any special attempts at reducing erosion is set at unity. Table 4-2 shows the fraction of erosion rates resulting when various other methods of management are used.

*Table 4-2. Management Factors*

| Management method | Management factor | Reduction of soil loss, % |
|---|---|---|
| *Residues* (corn stalks, straw, etc.) left on surface throughout the following year (2 tons or more per acre) | 0.50 | 50 |
| *Residues* moderately grazed but left on surface as above throughout the following year | 0.75 | 25 |
| *Residues* (corn stalks, straw, etc.) left on surface until planting time | 0.80 | 20 |
| *Mulch,* applied immediately after planting (manure, 6 tons or more, or crop residues, 2 tons or more per acre) (1 ton of manure contains about 500 lb of dry matter) | 0.60 | 40 |
| Winter *cover crops,* green manure or legume catch crops plowed at planting time (these crops have additional values for maintaining soil structure and organic matter) | 0.80 | 20 |
| *Plow-planting* (a minimum of cultivation and packing) leaves soil in condition to absorb rainfall | 0.60 | 40 |
| Use of *field cultivator* on meadow instead of plowing | 0.50 | 50 |

*S, the Slope-gradient Factor.* The percentage slope is raised to the 1.4 power to obtain the slope-steepness factor. This means that the amount of erosion is assumed to increase as the 1.4 power of the per cent steepness of the slope.

*L, the Slope-length Factor.* The length of slope, in feet, is raised to the 0.5 power to obtain the slope-length factor. This implies that the rate of erosion is proportional to the square root of the length of the slope.

*P, the Conservation-practice Factor.* The main items that are considered as conservation practices in this connection are contouring, strip cropping, and terracing. Others can, of course, also be included. Since the effectiveness of these methods in reducing soil losses depends on the steepness of the slope, the conservation-practice factors vary with the steepness of the slope, as is illustrated in Table 4-3.

*Table 4-3. Conservation Practice Factors*

| Slope group, % | Contouring factor | Strip-cropping factors | | Terracing factor† |
|---|---|---|---|---|
| | | *R-O-M-M* | *R-O-M-M-M-M** | |
| 1.1– 2.0 | 0.60 | 0.30 | .... | 0.30 |
| 2.1– 7.0 | 0.50 | 0.25 | .... | 0.25 |
| 7.1–12.0 | 0.60 | 0.30 | 0.20 | 0.30 |
| 12.1–18.0 | 0.80 | .... | 0.25 | 0.40 |
| 18.1–24.0 | 0.90 | .... | 0.30 | 0.45 |

* Rotation symbols: *R* = row crop
*O* = spring grain
*M* = meadow

† The terracing factors are based on the slope length between terraces that are properly designed for the given slope.

The lowest practice factors—indicating the greatest success from the practice—are found in the slope group from 2.1 to 7.0 per cent. On more gentle slopes the natural surface detention is quite effective by itself, while on the other hand, the detention created by the conservation practices decreases as the slope becomes steeper (see Fig. 3-30).

*E, the Soil-erodibility Factor.* Soil erodibility is the relative value at which a soil erodes. The soil-erodibility factor also reflects differences of rainfall intensity of the area in which it occurs. The erodibility of the Tama silt loam, an important soil in the western part of the corn belt and for which extensive experimental data are available, has been set at unity. It must be recognized that both the amount of previous erosion and the steepness of slope can alter the relative erodibility of various soils. For instance, a sandy soil is less erodible than a fine-textured soil on a gentle slope, while on a steep slope this may be reversed. Soil-erodibility factors are established for soils for which experimental data exist and for related soils in the same region.

Upon losing their original top layers, some soils erode at a faster rate and a few soils erode at a slower rate than originally, while others are hardly affected. These facts are reflected by the soil-erodibility factor, as illustrated for three soils in Table 4-4.

*Table 4-4. Factors for Previous Erosion*

| Degree of previous erosion | Soils | | |
|---|---|---|---|
| | Fayette | Clarence | Apkre |
| No, or slight, erosion.............. | 1.25 | 1.75 | 1.00 |
| ¼ to ¾ of original topsoil gone..... | 1.25 | 1.50 | 1.25 |
| Severely eroded.................... | 1.25 | 1.25 | 1.50 |

*Table 4-5. Recommended Rotations for Various Conditions of Steepness and Length of Slope: Degree of Erosion and Conservation Practices on the Miami Silt-loam Group of Soils in Indiana**

| Range of slope, % | Depth to subsoil, in. | Conservation practices | | | | | |
|---|---|---|---|---|---|---|---|
| | | No practices | Contouring | | Strip cropping | | Terracing |
| | | 0–200† | 0–200† | 200–300† | 0–200† | 200–300† | |
| 0–2 | Over 8 | 17 | 17 | 17 | 17 | 17 | 17 |
| 2.1–6 | Over 8 | 3 | 10 | ///‡ | 11 | 11 | 11 |
| | 3–8 | 2 | 9 | /// | 11 | 6 | 11 |
| | 0–3 or less | 1 | 3 | /// | 7 | 7 | 11 |
| 6.1–9 | Over 8 | 1 | 2 | /// | 7 | 3 | 11 |
| | 3–8 | 1 | 2 | /// | 3 | 2 | 6 |
| | 0–3 or less | 1 | 1 | /// | 2 | 2 | 3 |
| 9.1–12 | Over 8 | 1 | 2 | /// | 3 | 2 | 5 |
| | 3–8 | 1 | 2 | /// | 2 | 2 | 3 |
| | 0–3 or less | 1 | 1 | /// | 1 | 1 | 3 |
| 12.1–18 | Over 8 | /// | 1 | /// | 2 | /// | 2 |
| | 3–8 | /// | /// | /// | 1 | /// | 1 |
| | 0–3 or less | /// | /// | /// (Contouring not recommended on long slopes) | 1 | /// | 1 |

* The numbers in the columns indicate the rotations recommended for the various conditions.

† Length of slope, ft.

‡ ///, permanent vegetation recommended on all hachured areas.

Since it would be very cumbersome to calculate the amount of anticipated erosion for every individual case, tables have been prepared for the use of the farm planner. From these he can see at a glance whether a proposed combination of land use and management would result in a rate of erosion within permissible limits. An example is given as Table 4-5.

CROP ROTATIONS ARRANGED IN SEQUENCE OF INCREASING INTENSITY*

1. *g-M-M-M*
2. *R-g-M-M-M*
3. *R-g-M-M*
4. *R-g-g-M-M*
5. *R-g-M-M-R-g-*$g_{(sc)}$
6. *R-g-M*
7. *R-R-g-M-M*
8. *R-g-M-R-*$g_{(sc)}$
9. *R-g-g-M*
10. *R-R-g-g-M*
11. *R-R-g-M*
12. *R-R-g-M-*$R_{(sc)}$
13. *R-*$g_{(sc)}$
14. *R-R-R-g-M-M-R-*$g_{(sc)}$
15. *R-R-R-g-M*
16. *R-R-R-R-g-M*
17. *R-R-*$g_{(sc)}$
18. Continuous row crop

where *g* = small grain
*M* = meadow
*R* = row crop—corn, soybeans, etc.
(*sc*) = sweet-clover intercrop

This table is set up in such a fashion that the rate of soil loss *A* will be within permissible limits for a soil of given erodibility *E* and the actual amount of previous erosion. The variables considered are the length of slope *L*, the steepness of the slope S, and the specific erosion-control practices *P* used. An optimum level of management *M* is assumed to exist.

## POSSIBILITIES OF REDUCING THE EROSION HAZARD

Factors that combine to create the hazards of erosion in agriculture include soil, temperature, rain, runoff, slope, and wind. It is important to establish the extent to which these phenomena can be altered to decrease the erosion hazard. Table 4-6 shows these possibilities in a schematic form.

* From *Agron. Handbook*, Purdue University Agronomy Department, 1956.

*Table 4-6. Possibilities of Reducing the Erosion Hazard*

| Factors affecting soil erosion | Must be accepted | Can be altered |
|---|---|---|
| Soil. . . . . . . . . . . . . . | Soil type (texture) | Organic-matter content, aggregation, detachability, transportability, moisture content, productivity |
| Rain. . . . . . . . . . . . . | Intensity and amount of rain | Intensity with which rain strikes the ground |
| Wind. . . . . . . . . . . . | General wind condition | Wind velocity at ground level |
| Temperature. . . . . . | Air temperature | Extreme variations of soil temperature |
| Slope. . . . . . . . . . . . | Average steepness of slope, direction (aspect) of slope | Steepness of individual stretches of slope, length of slope, microtopography |
| Runoff. . . . . . . . . . . | . . . . . . . . . . . . . . . . . . . | Amount, rate, velocity, frequency, and place of runoff |

# 5

# *Methods of Soil Conservation*

## PROTECTING SOIL AGAINST EROSION BY WATER

Soil conservation is the maintenance of the productive capacity of the land while producing crops. Therefore the main soil-conservation measures are identical with ordinary methods of good agriculture. The outstanding ones are land use, tillage, plant nutrition, and water management. Under favorable conditions, these measures are sufficient. Frequently, however, specific soil-conservation techniques—so-called "supporting practices"—are required for satisfactory results. The line between ordinary farming methods and supporting practices is drawn here arbitrarily to include in supporting practices the methods that have gained general popularity only after the beginning of nationwide efforts to control soil erosion in the United States.

## GENERAL METHODS OF GOOD AGRICULTURE

### *Land Use*

While any type of land use affects soil productivity and soil erosion, only that land use is discussed here that is devoted to crop production. Residential, industrial, and traffic use of land will not be considered here. Some of these latter problems will be treated in Chap. 6.

In a previous chapter it was pointed out that the various crops have a different effect upon erosion. This is due to the nature of the crop plants and to the farming practices that are associated with them. The more erodible the soil, the more protective the land use should be.

**Land-use-capability Classes.** The U.S. Soil Conservation Service has devised a scheme of land-use-capability classification intended to make the best use of each piece of land without causing excessive erosion [32, 33]. It divides the land according to its capability into three main groups:

1. Land suitable for many uses—suitable for cultivation as well as for tree and forage crops
2. Land limited in use—suitable for cultivation with major restrictions and for tree and forage crops
3. Land very limited in use—generally suited only for permanent vegetation

These three main groups are subdivided into eight land-use-capability classes.

### LAND SUITABLE FOR MANY USES

Class I. Very good land that can be cultivated safely and easily with ordinary farming methods.

Class II. Land that can be cultivated safely with moderate conservation treatments. These soils may be slightly erodible or may have water or climate problems.

Class III. Soils with considerable limitations in use and that require intensive conservation treatments. Erosion, droughtiness, excessive wetness, overflow, or salinity may be the causes of the problem.

### LAND LIMITED IN USE

Class IV. Soils that are severely limited in use. They can be cultivated only occasionally and with extreme care. These soils may be erodible, droughty, wet, overflowed, or saline, so that the kinds of cultivated crops that can be grown as well as the number of years favorable for crop production are very limited.

### LAND VERY LIMITED IN USE

Class V. Nearly level land that is best suited to permanent vegetation. These soils are often stony, wet, subject to damaging overflow, or have a short growing season.

Class VI. Land that is suited for grazing or forestry, with minor limitations. These soils are usually steeply sloping; some class VI land may be severely eroded, shallow, wet, subject to damaging overflow, or droughty.

Class VII. Soils in this class are severely limited in use. The size of the conservation problems exceeds those in class VI. They may be steep, stony, shallow, droughty, wet, subject to damaging overflow, or eroded. These soils are best protected by natural vegetation and limited use.

Class VIII. Very steep and rocky or sandy or wet land. Useful for wildlife food and shelter areas or for recreational or water-yielding purposes. Not suited for commercial production of crops.

In addition to the land-use-capability units, capability subclasses have been established to indicate the major land-use problem, such as

Erosion and runoff, *e*

Excess water, *w*

Root-zone limitations, *s*

Climatic limitations, *c*

A system of symbols has been developed to show these limitations on the land-use-capability maps.

The capability classification is based on the physical limitations of the land that determine long-time profitable use. In addition to soil and climatic factors, present-day technology, economic conditions, and adapted crops are considered. Changes in these may eventually change the capability grouping of the land.

Naturally, the stress in this classification is on erosion control. The four groups of physical factors considered in the establishment of these classes are:

The permanence of the soil

The productivity of the soil

Interfering factors

Climatic environment

For a given area the soil type, the amount of erosion, and the topography determine the land-use capabilities. This classification indicates the most intensive system of cropping for a certain type of land that will keep soil losses within permissible limits. Table 5-1 illustrates that the more conserving the land use, the wider is the choice of land.

The concept of land-use capability is of great value in the planning for soil-conservation farming. This method is obviously not without faults. It is difficult to classify the land into the proper groups. Criteria to establish the boundaries between the different groups vary with the locality and with time and are influenced by the individual making the classification. Changes in tillage and fertilization methods as well as in soil deterioration and soil erosion have an effect on such a classification.

The greatest value of classifying the land according to its use capability is to indicate at a glance to the farmer and land planner the maximum intensity of agricultural use that can be practiced safely.

*Table 5-1. Land-use Possibilities According to Capability Classes*

| Examples of land use | Land-use-capability class | | | | | | | |
|---|---|---|---|---|---|---|---|---|
| | I | II | III | IV | V | VI | VII | VIII |
| Wildlife | Yes | Yes | Yes | Yes | Yes | Yes | Yes | Yes |
| Forest | Yes | Yes | Yes | Yes | Yes | Yes | Yes | No |
| Pasture | Yes | Yes | Yes | Yes | Yes | Yes | Maybe | No |
| Meadow | Yes | Yes | Yes | Yes | Yes | No | No | No |
| Small grain | Yes | Yes | Yes | Yes | No | No | No | No |
| Row crop | Yes | Yes | Yes | No | No | No | No | No |

In selecting crops for a given piece of land more than just soil erosion is considered. A droughty soil, for instance, fits a winter grain better than a spring crop. A poorly drained soil, even if no erosion hazard exists, should not be seeded to alfalfa, etc. An example of an erosion-hazard classification as a basis for land-use planning used in Germany is given below. This emphasizes the importance that is accorded preceding and potential erosion in determining land-use capability.

EROSION-HAZARD CLASSIFICATION ACCORDING TO JACOBUS LÜTTMER*

I. Soils that are friable and permeable to a depth of 60 centimeters. They must have had little or no previous erosion and may occur on slopes up to 7 per cent.

II. Soils on slopes up to 12 per cent (in draws up to 10 per cent) on which slight to moderate erosion has occurred, but at least 30 cm of friable soil remains.

III*a*. Soils on slopes up to 22 per cent (in draws up to 15 per cent)

* Jacobus Lüttmer, Bodenschutz in der Flurbereinigung, *Schriftenreihe fuer Flurbereinigung, Bundesministerium fuer Ernaehr., Landwirtsch. und Forsten,* Heft 14, 1957.

on which less than 30 cm of friable material remains. These soils have considerable erosion hazard because of the shallow soil profile.

III*b*. Soils on slopes up to 32 per cent (in draws up to 22 per cent). The erosion hazard is serious, and it can be used for cropping only where intensive conservation measures are employed.

IV. Soils which are subject to frequent overflow or which occur on slopes greater than 32 per cent (22 per cent in draws). Agricultural use is impossible. They must be kept under permanent vegetation, such as dense sod or a thick stand of trees.

**Pasture and Forest Management.** Experience and numerous experiments have shown that pasture and forest protect the soil from erosion better than other forms of land use. This is because both represent a form of permanent vegetation. The soil is not tilled, and it is constantly covered by a protective mantle of plants and plant residues. The value of pasture and forest is recognized by the fact that the land-use-capability classification of the U.S. Soil Conservation Service assigns the two most erosive classes of cropable land, class VI and class VII, exclusively to these forms of land use.

It must be recognized, however, that pasture and forest can only control erosion effectively if they are properly managed. In the case of pastures this means establishment of a dense sward by selection of the proper species, fertilization, and protection from overgrazing and from grazing when the soil is wet. The dense sward is needed to protect the soil surface from the direct impact of the raindrops and to supply a large mass of roots to maintain an ample infiltration capacity. Grazing when the soil is wet and in a plastic condition causes it to be compacted unduly, so that the larger pores are closed and the rate of infiltration is greatly restricted. In severe cases the hoofs of the animals may actually push the plants into the ground and expose the soil, so that it can easily erode. Erosion becomes a pronounced hazard in a pasture at places where the animals tend to make paths. The constantly repeated pressure and the excessive grazing can easily result in severe gullying. Such locations require special protection, principally by diverting the animals.

Where pastures have been damaged and erosion is apparent, fertilization, reseeding, mulching, and temporary protection from grazing are needed. Sometimes smoothing out of the land surface may have to precede these operations.

The value of forest for soil and water conservation has been emphasized in a previous chapter. The benefits derived are greatest from a forest in its natural condition. The cutting of trees in itself does not decrease greatly the soil-conservation value of the forest. The removal of timber and other forest products, however, represents a definite erosion hazard. Snaking out logs with large tractors can result in deep cuts in the soil, particularly because the ground during timber harvest is frequently wet. Also, truck roads used in the forest can develop into gullies unless they are carefully planned and maintained. Other erosion hazards in the forest are fire and grazing. A severe fire that destroys both the ground cover of plants, mulch, and the trees as well can result in disastrous erosion in a mountainous area. But even a ground fire that leaves the larger trees unharmed removes much of the protection from the soil and exposes it to rainfall and the eroding force of surface runoff. The destruction of organic matter by fire, especially if it is repeated regularly, causes the soil structure to deteriorate and to make the soil less pervious to water.

Grazing is another cause of creating erosion hazards in the forest. Since the trees filter out the sunlight, there is little forage in a forest, and the animals are forced to roam about, thus trampling the originally loose forest floor and decreasing porosity and infiltration capacity. Moreover, grazing destroys the young forest plants and, if severe enough, may prevent reproduction. Grazing has no place in a woods, certainly not in the dense woods of the humid climates. In dry climates, where trees stand farther apart and grassy undergrowth exists, limited grazing may be expedient, especially if careful attention is paid to a maintenance of a full ground cover. But even under this condition grazing is damaging to reproduction.

From the viewpoint of soil conservation it is best never to level a forest completely, but to manage it on a sustained-yield basis. This implies cutting mature trees at regular intervals and allowing the smaller ones to remain, thus always maintaining a cover. Even where clear-cutting is practiced, it is well to allow a few trees of desirable species to remain as a source of seed.

Fields that are too steep for cultivated crops or pasture should be reforested. Since tillage would expose soil to erosion, it is best to plant the young trees with the minimum of ground preparation. Usually a hole dug with a mattock or a slit opened by a tree-planting machine suffices. It is important to select species that will grow quickly and get along well on the site. In the American Middle West and South various species of pine trees have served well as pioneer trees on the

**Fig. 5-1.** Cover crop in corn. Crotelaria in this Alabama corn field helps to protect the ground from erosion and adds organic substance and nitrogen to the soil. (*USDA*)

eroded fields of plantations because of their small nutrient and moisture requirements. In some cases the establishment and growth of forest trees has been speeded up materially through fertilization. As the relative value of forest products increases, the use of fertilizer in the woods will probably increase. Some trees, especially pines, are very sensitive to fertilizer near the roots. Therefore care has to be taken in its application.

**Choice of a Cropping System.** The establishment of the proper land use for every area on the farm does not in itself determine what crops to plant. This will depend also on climate, economic conditions, and the personal preference of the farmer. In the case of the first three land-use classes, the choice of cropping systems is large and frequently difficult. The greater the erosion hazard, the smaller will be the proportion of row crops and the greater will be the proportion of sod crops. Methods have been worked out to estimate the amounts of

**Fig. 5-2.** Cover crop in an orchard. Rye is used in this New Jersey orchard to protect and improve the soil. It will be disked up soon to incorporate it into the soil and to avoid excessive moisture loss by transportation. (*USDA*)

erosion as they might occur under various conditions of soil type, steepness and length of slope, rotations, and specific soil-conservation practices. Such estimates will permit the choice of a combination of these factors that will restrict erosion to "permissible" rates.

The amount of erosion that can be expected under a given cropping system can be considerably reduced by the use of cover crops. Grasses, small grains, or legumes seeded between row crops or during an otherwise fallow period reduce the erosion hazard to almost that of a meadow. While establishing an intercrop is not always simple, wide row planting of the main crop, heavy fertilization and the choice of the most adapted species are some of the means of ensuring success of this practice.

On land suited for limited cultivation the choice of a cropping system is restricted to small grain and sod crops. Usually meadow is maintained as long as it gives satisfactory yields. The land is then plowed or only cultivated, small grain is seeded, and a new meadow established in the grain.

In case of the four land-use classes not suited for cultivation the only crops are pasture plants, forest trees, or shrubs. The climatic and soil conditions will usually determine which species should be used.

## *Tillage*

**Purposes of Tillage.** The main purposes of tillage are to prepare a seedbed, to prepare a root bed, and to eliminate competing vegetation.

*To Prepare a Seedbed.* This means to condition the soil surface so that the seeds of the crop plants can readily germinate and take root and the top can break through the soil surface with ease. Different crops have different requirements in this respect. Potatoes manage quite well in a rough soil, corn has to have a somewhat finer structure, small grains have still higher requirements, while grasses and small seeded legumes are most exacting. Generally speaking, the larger the seed or the seed piece, the less preparation is necessary.

*To Prepare a Root Bed.* Once the seed is germinated, the roots need a favorable habitat. This includes sufficient looseness to penetrate the ground easily, sufficient consistency to hold the roots and to conduct water, enough porosity for adequate aeration, and a biologic activity sufficient to mineralize plant-nutrient elements at the rate they are needed.

*To Eliminate Competing Vegetation.* This includes both weeds and other crop plants. Where perennials or biennials precede another crop, it is usually necessary to kill the previous crop through tillage.

Tillage may have other purposes: removal of dead plant residues from the surface, making the soil more absorptive for air or water, moisture retention, disposal of excess water, incorporation of green or barnyard manure, etc.

**Effects of Tillage.** In pursuit of the first two purposes of tillage, we are striving for a soil in *good tilth*, not merely a temporarily friable soil. Good tilth may be defined as the condition of the soil in which many aggregates of desirable dimensions exist in water-stable form.

Most methods of tillage decrease the size of aggregates and encourage the oxidation of organic matter, which means a loss of bacterial food and a decrease of aggregate-binding material. Every tillage has a tendency to press some of the soil together, even though much of it may be loosened up. The weight of the implement and the tractor account for part of this.

Therefore, while tillage usually results in a temporary increase of pore space, better aeration, and a spasmodic activation of the microbes, it has a long-time tendency to decrease the organic-matter content and the microbe numbers and to make the soil more compact, more detachable, more transportable, and hence more erodible.

Tillage is needed for the production of most crops, but from the viewpoint of soil conservation, it must be considered a necessary evil. To conserve the productivity of the soil, it is well to till no more than necessary to accomplish the principal aims: to get the soil into desirable physical condition and to discourage competing vegetation.

**Contribution of Individual Tillage Implements.** The action of some of the more important tillage implements deserves attention.

*Plow.* The plow is an excellent tool to break up the soil into large clods. The lifting and turning causes the soil to be separated at the natural cleavage planes. Therefore the plow does not destroy soil structure as much as tillage implements do that use knives to cut the soil (e.g., the disk) or that use high speed to pulverize it (e.g., the rototiller) [36]. Plowing covers all crop residues and vegetation. It makes the ground more absorptive for rain, but usually this advantage lasts only a month or two. Like all other tillage operations plowing gives the best results when it is done on soil of intermediate moisture condition (pF 3.0 to 4.0). If the soil is too wet, it puddles; if it is too dry, it breaks into large chunks and powder. The plow tends to trowel the plow sole into a tight condition (plow pan), especially if the shares are dull [44, 50].

If the plowing is done in the fall in a climate where much freezing and thawing occur during the winter, the aggregates of the unprotected soil are largely broken up by the frost, making them susceptible to erosion. Dead furrows frequently serve to collect runoff water and start gullies.

Under many conditions plowing alone is sufficient to prepare the ground to plant corn or other row crops, provided the soil in the row is leveled and firmed in the planting operation. This "minimum tillage" has the advantage of leaving the soil loose and receptive for rain water. It discourages weed growth and is economical of time and labor. This system was developed in Michigan [16], and it is used in many other states.

*Disk.* The disk is a tillage implement of quite general value. By using different types of disks and different settings, a large number of tillage requirements can be met. From the point of view of soil conservation, the advantage of the disk is that it can be used on land that is covered with crop residues. Probably the worst feature of the disk is that it cuts through natural soil aggregates and exposes the internal parts that are not sufficiently protected by organic and microbial substances. Therefore rain and frost soon will break down soil that has been disked and give it a smooth, tight surface, unless it is protected by mulch. Usually several trips of the disk over the ground are needed to smooth out the surface soil. This compresses the lower part of the plow layer, while the surface soil becomes loose. Thus the disk may produce a good seedbed but a poor root bed.

*Cultivator.* Cultivators serve to loosen the soil without turning it over. They leave crop residues largely uncovered and can therefore be used for mulch tillage. Cultivators can penetrate to a depth of a foot or even more. When they are used at favorable soil-moisture conditions, they prepare an excellent seedbed and root bed. Since cultivators mix the layers of the soil only to a limited extent, special consideration has to be given to the placement of fertilizer, if no plow is used. Mounting a hopper on the cultivator and placing the fertilizer spouts behind the cultivator chisels or sweeps put the fertilizer deep enough into the ground.

*Surface Tiller.* A variety of implements is used to till the soil just under the surface. Rod weeders and V-shaped sweeps go through the ground at 2- or 3-inches depth and cut off the roots of weeds or of crop plants. They are of particular value in dry climates, because the residues are left on the surface for protection of the soil and the plants that are cut off do not start growing again as they frequently do on moist soil.

*Subsoiler.* The chisel or subsoiler opens tight soil deeper than any of the other tillage implements. It facilitates percolation of rain water and root penetration. A lasting benefit of subsoiling can only be derived if the roots permeate the newly created openings. Application of lime and fertilizer help to achieve this [18, 39]. When subsoils are dense and low in fertility, subsoiling in combination with fertilization of the subsoil frequently results in crop yield increases. Subsoiling can be made very effective by blowing chopped crop residues into the chisel groove [52]. This operation, called "vertical mulching," prevents the groove from closing up and ensures rapid intake of rain water. If subsoiling and vertical mulching are done on the contour, they are very effective in runoff control.

**Role of Tillage in Soil Conservation.** If we exclude contour tillage, tillage as such is of little direct advantage in soil conservation. By loosening the ground, it may make the soil more absorptive for rain water, thus decreasing runoff. But this benefit is only temporary. Moreover, loose soil is highly erodible. Its main value in this connection is to help to establish vigorous crop plants.

To achieve the main purpose of tillage and at the same time to conserve the soil, the following suggestions are offered:

Till no more than necessary.

Till only when the soil moisture is in the favorable, intermediate range.

Use chemical weed killers where practical.

Plow erodible soils in the spring.

Vary the depth of plowing.

## *Plant Nutrition*

**Role of Plant Nutrition in Soil Conservation.** A vigorous growth of crop plants is the best insurance against excessive erosion. The direct protection against rain-drop impact and surface washing provided by the plants is matched in importance by the improvement of soil structure resulting from the addition of organic matter of both roots and tops. An ample supply of plant nutrients in the soil is one of the main requirements of a strong and healthy crop.

**Replacement of Lost Plant Nutrients.** There is a continuous drain on the plant nutrients of the soil. They are lost in the harvested crop parts, by leaching, by erosion, and by chemical fixation. In some cases nitrogen and sulfur may be lost by evaporation. These losses are compensated to a certain degree through nitrogen fixation by symbiotic and nonsymbiotic microbes, through decomposition of plant residues, through manure, through weathering of the soil minerals, and through rain and dust. On alluvial soils flood water may add substantially to the plant-nutrient supply. In general, however, the natural additions to the supply of fertility on the field under conditions of most types of agriculture are smaller than the losses, and therefore systems of fertilization with manure, lime, and commercial fertilizer have been adopted in most humid or irrigated areas.

**Fertilizer Requirements.** To tell the amounts of plant nutrients required for a given field, various methods can be employed.

*Field Experiment.* A field experiment using various combinations of plant nutrients is the safest method to determine the fertilizer requirements of a given area. In order to make such an experiment acceptable to the farmer it must be very simple, preferably testing for only one element. More elaborate experiments are needed to determine the required amounts of the various nutrients. One of the disadvantages of field experiments is that the information resulting from them cannot be applied immediately, but only to ensuing crops.

*Balance Sheet.* A balance sheet of the gains and losses can be set up and the differences made up by use of commercial fertilizers. In doing this, recognition has to be given to possible desirable yield increases over the present levels and to the fact that only part of the applied fertilizer becomes available to the crop. It is generally assumed that only one-half the applied nitrogen and potash and one-tenth the applied phosphorus is taken up by the first crop.

*Soil Tests.* Chemical tests to determine the fertilizer needs of the soil are being used in increasing numbers from year to year. Many of these tests have been found to give an excellent picture of the fertility status of the soil as determined by field experiments. Usually they are applicable over a region of similar soils and climate. None of the testing methods developed has proved satisfactory for all soils, all locations, and all crops. As a matter of fact, none of them gives an absolute fertilizer requirement for a given case. There are at least two reasons for this. Even the most elaborate chemical methods cannot duplicate the absorption of the nutrients by plant roots. The roots take several months for this process, while chemical extracting solutions come in contact with the soil only hours or even minutes. Secondly, the plant-nutrient requirement depends upon available moisture and temperature, and it is impossible to predict the weather of the coming season. Nevertheless, it has been found that some test methods give good correlation with field response for specific locations and conditions. Elsewhere, other tests are more suitable. The degree of reliability varies. In most cases the results for the individual plant-nutrient tests are subdivided into five or even only three levels of availability, for instance, "high," "medium," and "low." Probably the most important single factor in the successful use of soil tests is correct interpretation.

*Plant Tests.* Either the composition of the leaves or other parts of the plant may be determined quantitatively, or the composition of the plant sap may be tested semiquantitatively for the important plant nutrients. Especially the latter technique has considerable value in determining the most deficient of the main plant nutrients. Since it can be performed in the field, it has become quite popular and useful. Neither this technique nor quantitative foliar diagnosis permits the determination of the actual amounts of plant nutrients needed.

*Carbon-Nitrogen Ratio.* Addition to the soil of carbonaceous crop residues such as grain straw, timothy sod, or corn stover temporarily widens the carbon-nitrogen ratio and results in a largely expanded microbial activity and an acute nitrogen tie-up. If this coincides with a large nitrogen requirement of the crop, a corresponding application of nitrogen fertilizer is advisable. The method of incorporation of this organic matter has a pronounced effect on the nitrogen requirement. Organic residue placed on the plow sole decomposes much slower than that dispersed through the upper layer of the soil with the disk, and therefore less nitrogen is needed where the plow is used than where mulch tillage is practiced.

*Experience.* One of the safest guides to determine the plant-nutrient requirements of a farm is close observation of the crops. The appearance of the plants frequently indicates whether they are well supplied or whether they are starved for any specific nutrient. Such observations are of greater value still if different combinations of plant nutrients have been used on different parts of the same field. The size of the yield is a useful clue of the adequacy of fertilization. The occurrence of indicator plants (usually weeds) helps also in this respect. If manure spots show up plainly in the growth response, the field needs fertilization.

**Systems of Fertilization.** Once the plant-nutrient requirements of a farm are recognized, a system of fertilization must be developed. Economic, physical, historic, and human factors enter into such a decision. The higher the relative price the farmer receives for the product, the higher is the level to which the field will be fertilized. Land near the farmstead and land of smooth topography will generally receive more manure and more fertilizer than land farther away and on the steeper slopes. The outstanding factor, however, that determines the system of fertilization is the knowledge and the business acumen of the farmer.

In planning the best method of fertilization for a specific case many factors must be considered even after the plant-nutrient level of the soil has been determined and the nutrient requirements are known. Should only the losses be replaced, or should the nutrient level be increased? It is axiomatic that the nutrient element that is available at the lowest price—in terms of its effect on yield increases—is used at the highest rate, in an effort to allow the other more expensive fertilizer nutrients to be fully effective. Even though the purchase prices are known, it is not always simple to determine which nutrient will pay best for itself. On acid land in an area where limestone is plentiful, the soil should be amply limed before other fertilizers are used. In some parts of the world, lime is so expensive that liming to near neutrality (pH 6.3 to 6.6) is prohibitive. Where proximity to mines or other sources makes phosphate or potassium inexpensive, they will be used in large quantities. The constantly improved methods of conversion of atmospheric nitrogen to fertilizers in solid, liquid, and gaseous form have lowered the price of this nutrient in comparison to the others to a point where nitrogen frequently is the fertilizer that gives the highest return on the investment.

Obviously, the price alone is not the criterion of the relative amounts of the individual nutrients to be applied. The farmer attempts to bring about "balanced" nutrition of his crops. Just what represents balance of plant nutrients has been the subject of many investigations without providing a universally applicable answer. Essentially, it is a combination of nutrients that will result in the best possible growth of the plant under the given climatic conditions. It is believed that balanced nutrition can best be achieved by an ample, but not excessive, application of all needed nutrients to the soil.

Another important problem in developing a system of fertilization is the matter of distribution. Most authorities agree that a combination of broadcasting and of starter fertilization is generally the best. The question is: How should the amounts of nutrients be distributed between these two methods? The more intensive farming becomes, the more important it is that plant roots find nutrients throughout the soil, and the more must broadcast fertilization be favored. To what depth soil should be fertilized is a problem that has been inadequately investigated so far. It seems reasonable to expect that plant roots will grow profusely as deep as ample nutrition and aeration are available. Application of fertilizer to a depth greater than previou ly used should result in deepening the layer permeated by many roots and consequently in better physical condition. A more rapid intake of water and consequently diminished runoff are the advantages the soil conservationist would expect from deep fertilization.

The main plant-nutrient elements can be applied in a variety of chemical and physical forms. The degree of solubility and the extent to which the fertilizer ions are held by the soil are the chemical variables. Fertilizers may be in solid or liquid form. Nitrogen is also applied as ammonia gas. Solid fertilizers vary in granulation. In recent years, most solid fertilizers have been carefully prepared to run freely through the fertilizer spreader. They are either formed into small pellets and protected from hygroscopic moisture by a coating or some inert material is added to prevent caking.

**Mutual Relationship between Soil Conservation and Fertilization.** It is of importance to realize that soil conservation and fertilization are mutually beneficial. As has been pointed out, adequate and even ample fertilization is a necessary component of a sound soil-conservation system. But conversely, the protection of the soil from physical deterioration and from erosion is a prerequisite of a successful program of fertilization. Fertilizer that is swept off the field or that is placed on a tight, poorly aerated soil is an economic loss.

## *Water Management*

If we want to regulate soil water by ordinary good farming methods with the purpose of controlling erosion, we have to attend to the following items:

Cut down the amount of runoff.

Decrease the velocity of runoff.

Regulate the soil moisture in such a way that the soil is least transportable by wind or water.

Regulate the soil moisture so that microbes and crop plants find their optimum moisture and aeration conditions.

Make the existing soil moisture most effective.

Fortunately, these requirements are closely interrelated and can be achieved simultaneously.

The intermediate soil-moisture content, the moist range (pF 3.0 to 4.0), represents the optimum condition for crop plant growth and microbial activity, because satisfactory amounts of both water and air are present. As an abundant growth of microbes and plant roots is the prime requirement of good tilth, this moisture content is also best for making the soil porous and resistant to detachment and removal. The high rate of infiltration resulting from a porous, water-stable structure and the resistance to surface flow by the dense plant cover help to decrease the amount and the velocity of runoff.

From this it evolves that water management for soil conservation has to aim at the intermediate moisture condition. Or in other words, we have to avoid the extremes of soil moisture. How can this be done in practical agriculture?

### TO KEEP THE SOIL FROM BEING TOO WET

Improve soil structure so that internal drainage can proceed unhampered.

Where necessary, use tile drains and ditches to lower the water level.

Avoid accumulation of moisture in the surface by freezing.

Use insulation (crops and crop residues).

Make use of plant transpiration.

Reduce the amount of rain water that runs into the low places by increasing as much as possible the infiltration capacity of the surrounding land.

TO KEEP THE SOIL FROM BEING TOO DRY

Improve soil structure so that infiltration can proceed unhampered.

Increase the water-holding capacity of the soil.

Increase the depth of root penetration to give the crops a bigger feeding zone by drainage and by opening the subsoil with legumes or subsoilers.

Do not plant crops too thickly.

Control weeds.

Decrease evaporation by keeping the soils shaded, and do not stir them too often.

Prevent surface runoff water from becoming channelized, and make sure that it spreads over a wide area.

### *Organic-matter Management*

The value of organic matter in the soil is generally recognized. It supplies energy and carbon for microbes. It increases aggregation of soil and helps to make aggregates water stable. Consequently, it increases porosity and aeration and therefore increases infiltration and percolation capacity. This in turn reduces runoff and decreases the erosion hazard. As a surface cover; or mulch, it protects the soil from the impact of raindrops and from the wind. In addition, decomposition of organic matter mineralizes plant nutrients, thus fertilizing the soil. The high organic-matter content in virgin-forest and prairie soil is part of the reason why these have little surface runoff and practically no erosion. In order to derive the full benefit from organic crop residues on agricultural land it is necessary to develop the best method of management for the individual case.

The first thing is to produce large amounts of high-quality organic matter. Selection of adapted crop varieties, proper tillage, and ample fertilization are the main requirements. Slow oxidation of the residues results in relatively large amounts of the intermediate decomposition products that are active in cementing together soil particles. When the residues oxidize quickly, they are soon broken down to carbon dioxide and water. These have no beneficial effect upon soil structure. Therefore residues should be managed in such a way that they decompose slowly, as far as that is possible. The roots of the crops are ideally placed for this purpose. They are well distributed throughout the soil, and yet they are protected from excessive oxidation. The incorporation of the above-ground portions presents a problem. As long as they are not in contact with the soil—as is the case with stubble or to a certain extent with corn stover after picking—decomposition is very slow. Plowing under covers up the residues and starts decomposition. The rate is determined by the temperature and by the moisture content of the soil and the distribution of the residues in the soil. If crop residues are bunched together on the bottom of the plow sole of a wet, tight soil, the oxygen supply may be so limited that undesirable carbonization results. The fastest decomposition results when the residues are well distributed near the surface in a moist soil. The most nearly ideal (intermediate) rate of decomposition is brought about by distributing the residues evenly throughout the plow layer. When crop residues are used as mulch, the rate of decomposition depends on the degree of mixing with the soil. It is the slowest when there is least mixing.

Organic matter in the soil is of transient nature. Each year new material is added, and organic matter continuously decomposes. Under most conditions gains and losses are balanced, and there are little if any changes in the total amount of organic matter. It is difficult to increase the store of organic matter, even if extra amounts are regularly added. This may not be so serious, because soil derives the main benefit from organic matter during the first period of decomposition. Nevertheless, it is desirable to build up the store of humus by bringing about conditions favorable for production of large amounts of residues, especially roots, for leisurely decomposition and for the protection of soil from erosion.

## SPECIFIC SOIL-CONSERVATION TECHNIQUES

### *Mulching*

**Purpose of Mulching.** Proper land use, fertilization, contour cultivation, terracing, and strip cropping help to maintain desirable soil structure, permit ready drainage, and protect the soil from erosion. But during plowing and until the first cultivation, soils are endangered in spite of these measures. In areas of irregular slopes contouring and terracing are not feasible. In such areas direct protection of the soil by surface cover and by increased stability of the soil aggregates is needed.

Covering the soil with crop-residue mulches increases the infiltration capacity and decreases runoff and erosion losses in practically all cases. The desirability of mulch for erosion control is generally well recognized. The points that need more careful attention are these:

Which crops require mulch for erosion protection?

What are the sources of mulch?

Should and can the crop residues be left on top of the ground, or should they be partially incorporated in the soil?

How much crop residue is required for an effective mulch?

Will mulches interfere with cultivation?

How do mulches affect weed growth?

How do mulches affect crop yields?

How to fertilize when mulch tillage is practiced?

What type of equipment is needed?

**Effects of Mulching.** Mulching affects the physical, chemical, and biological conditions of the soil. While the over-all result is good soil conservation, there are various points that are disadvantageous. Both sides have to be weighed carefully before mulch culture is adopted. The physical, chemical, and biological effects are enumerated below.

When a bare soil surface is covered with a mulch of fresh straw, physical, chemical, and biological phenomena will occur as a result. Comparing the mulched with the unmulched soil we can enumerate these phenomena as follows:

PHYSICAL

Less direct impact of raindrops
Decreased amount and distance of splash [24]
Less dispersion of surface soil, less crusting [43]
Less internal erosion (clogging up of pores)
Smaller fluctuations in soil moisture and soil temperature
Lower temperatures in spring and summer
Higher temperatures in winter [8]
More gentle vapor-pressure gradients in the soil surface
Decreased frequency of freezing of soil

Decreased rate of freezing of soil

Less accumulation of ice and water in the soil surface

Decreased depth of frost penetration (Fig. 5-3)

Increased aggregation of surface soil, improved soil structure [41]

Greater resistance to detachment by wind and water

Greater porosity, increased water-holding capacity (mulch itself holds very little water) [6]

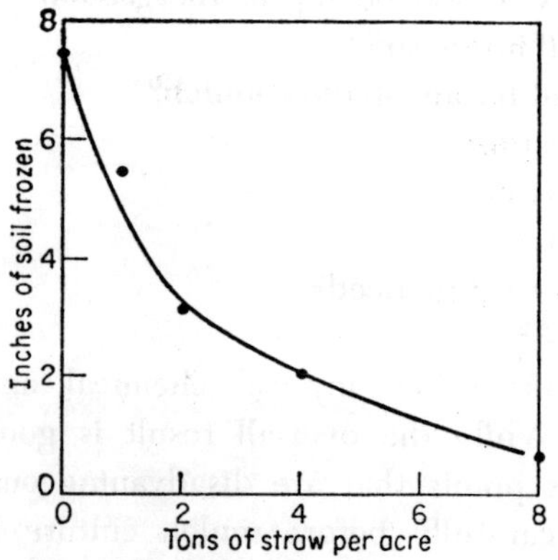


**Fig. 5-3.** Frost protection by straw mulch. Eight tons of straw per acre nearly prevented the soil from freezing during a severe winter in Indiana.

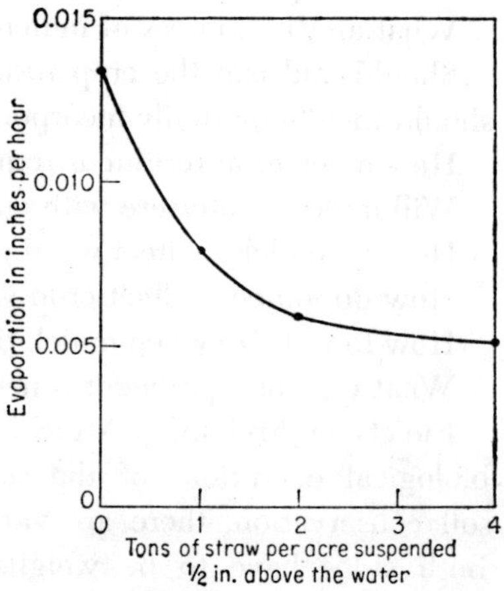


**Fig. 5-4.** Mulch reduces evaporation. Two tons of straw per acre above small containers filled with water are sufficient to reduce evaporation to less than half the amount of unprotected water.

Higher infiltration capacity, increased amount of percolation [38]

Water will not stand as long on the surface

Less runoff and water erosion [3, 21, 45]

Less evaporation (Fig. 5-4)

Better retention of microtopography, especially during the winter

Decreased wind velocity just above the ground

Less wind erosion

CHEMICAL

Tie-up of available nitrogen in organic form shortly after application of straw

Reduction of surface soil due to readily decomposing carbohydrates

Release of organic nitrogen in available form later on

Possible inducement of increased nitrogen fixation

Temporary tie-up of phosphorus in similar manner to nitrogen tie-up

Potassium is made less available because of reduced soil conditions

Increase of total potassium (fresh straw is rich in potassium, 0.5 to 1.0 per cent $K_2O$)

Less potassium made unavailable because of less wetting and drying

More loss of plant nutrients in percolate because of increased percolation

Less loss of plant nutrients in runoff because of decreased amounts of runoff

BIOLOGICAL

Increased microbial activity near the surface because of increased energy supply and because of more uniform moisture and temperature conditions [27]

Larger populations of insects, including crop pests as well as their enemies, worms and rodents

Easier establishment of most plants, especially the small seeded ones, including weeds

Effects on crop yields depend on the type of crop, the quality and thickness of the mulch, the nutrient supply of the soil, and the texture and the structure of the soil.

**Place of Mulch in Rotation.** Mulches as means of erosion control are of value for those periods in which the crop itself does not adequately protect the soil. This includes row crops, small grains, meadow seedings, and fallow. Experience shows that mulch tillage is more needed for row-crop and fallow periods but is more easily applied and used on small grains and meadow seedings, because these do not normally require cultivation. Mulch is also used successfully on truck crops, orchards, and even in the establishment of young forest trees under particularly adverse conditions. Land-use-capability classes II, III, and IV are those where mulch tillage is used most.

**Sources of Mulch.** The natural sources of mulch are agricultural by-products, for instance, straw, stubble, stover, corn cobs, meadow residues, and manure. Sawdust, wood chips [40], shavings, paper, or plastic film are occasionally used as mulch [38]. In agricultural practices it is best to "grow the mulch in place," that is, to use residues in the same field where they grew.

Small grain produces much mulch at a time when not much mulch is needed. The straw may be removed from a grain field at harvest time. Clipping the stubble with a mower will generally provide an ample amount of mulch for erosion protection. Frequently, small grain is seeded into a plowed field and requires a special mulch application. In this case straw or, better yet, manure can be spread with a manure spreader.

**Methods of Application.** Mulch may be used exclusively as a soil cover, or it may be partially mixed with the ground [7, 8]. As a cover it is more effective in protecting the soil from the direct impact of the rain drops. However, if it is partially mixed with the soil surface, it decomposes sooner and helps to make the soil more rapidly detachment resistant [42]. This latter method is not using the residues strictly as "mulch," but it is included here because it is a valuable method of erosion control and it is practically impossible to make a clear distinction between the two methods.

**Rate of Application.** The rate of mulch application in the field should be thick enough to cover two-thirds to three-quarters of the soil from view. This will suffice to protect the soil from much of the impact of rain and practically stop soil transportation by splash. It reduces appreciably evaporation, temperature fluctuations [59], and freezing and thawing of the soil. On the other hand, this quantity of mulch does not suppress the growth of most plants and permits the atmosphere to reach the soil sufficiently for adequate aeration. This quantity corresponds to 1½ to 2 tons of straw or 4 to 5 tons of barnyard manure per acre.

For certain horticultural purposes mulch should be applied at considerably heavier rates for weed control (e.g., around orchard trees). A thick layer of mulch checks the growth of plants; a thin one encourages it. It is seldom advisable to use mulches in excess of 2 tons dry material per acre over a larger area, because this would absorb an excessive portion of the rain water. Much of this evaporates and is lost to plant growth.

**Time of Mulching.** Many of the mulching materials are highly carbonaceous residues with relatively small amounts of nitrogen. Shortly after the application of such mulches a great demand is placed on the nitrogen in the soil for the decomposition of these residues. If this period coincides with the most active growth of the crop, the dearth of mineral nitrogen causes reduced crop growth. It is wise, therefore, to allow the residues to decompose partially before the crop is planted. This decomposition may proceed previous to application, as is frequently the case with manure or compost, or it may go on in the field several months before the principal growth of the crop starts. This is accomplished, for instance, by the application of straw on winter-wheat land just after the wheat is seeded. By the time the wheat requires much nitrogen in the late spring, a good share of the easily decomposable organic matter is already destroyed. Maceration of corn stalks before winter is another example of mulching in advance of planting.

Beside this chemical consideration the physical conditions also speak for an early application of mulch. If the ground is bare at the time of seeding a crop, mulch should be applied immediately to protect the soil from erosion. And even if we disregard the immediate erosion danger, early mulching proves advantageous, since the structure improves progressively after the mulch is applied.

**Problems of Mulch Tillage.** In many cases mulch fits very well into ordinary farming methods; in fact, it has long been an accepted practice. Manuring winter wheat, mulching strawberries (their very name points to the use of straw as mulch), and mulching lawn seedings and orchard trees are examples. Some minor problems arise sometimes even in these cases, problems such as increased weed growth, but the major difficulties occur where mulch is used with cultivated crops. Tillage of a meadow in preparation of a seedbed for corn is not simple if a large proportion of the meadow residue is to stay on the surface. Planting corn or soybeans below a mulch requires special care or even special equipment. Due to the faster decomposition of organic residues near the soil surface than when they are plowed under, plant nutrients may be temporarily unavailable to crop plants, and therefore an extra fertilization, especially with nitrogen, may be needed [27]. Once a soil is built up to a fair degree of fertility, this measure becomes less important.

Since mulch tillage has the tendency to improve the surface soil, turning the soil with a plow once in several years has the advantage of deepening the enriched layer of the soil.

**Fig. 5-5.** Stubble mulching. Mulch tillage in this wheat stubble field is accomplished by the use of V-shaped sweeps. (*USDA*)

**Mulch-tillage Implements.** Only implements that leave most of the crop residues on the surface of the ground can be used for mulch tillage. The disk cuts up the residues and loosens the surface soil. V-shaped sweeps, straight horizontal blades, the rod weeder, and the rotary hoe are other implements that break up the surface soil without turning it over (Fig. 5-5). The very necessary loosening of deeper soil layers is accomplished by cultivators. They can also serve to place fertilizer deep into the soil by attaching spouts behind the cultivator shanks. The fertilizer is fed from a hopper placed on the cultivator.

After the ground has been prepared with such implements, seeding or planting presents no difficult problems. It can be done with conventional equipment. A special machine has been designed that mulches the ground, fertilizes it, and plants corn—all in one operation. Cultivation of mulch-planted row crops can usually be accomplished without specialized equipment when the mulch pieces are not too long. Otherwise, disk cultivators can be used. Much of the cultivation work can be saved by the proper use of herbicides.

**Mulch Tillage and Crop Yields.** While mulching the ground either by covering it with crop residues or by mulch tillage protects the soil from erosion and helps to maintain a favorable moisture content, it may sometimes result in yield decreases. This is true especially on soils of medium and fine texture in humid sections. One of the reasons for this is that residues on or near the surface of the soil are rapidly decomposed and use up much oxygen. The lack of oxygen and the accumulation of carbon dioxide in the ground are enhanced by high moisture content, which reduces the air space. In addition, the microbes that break up the residues require large amounts of nutrients in available form, especially nitrogen. This temporary tie-up of nutrients and the unfavorable aeration conditions together frequently cause lowered yields of nonleguminous crops planted in mulched ground. To get high yields in such locations under mulch culture the following considerations with regard to aeration and plant nutrients must be given special attention:

The main period of decomposition of the residues should not coincide with the main growth of the crop; it is best if it precedes it.

The soil should be loosened deeply and kept loose so that aeration can proceed unhampered.

A high fertility level must be maintained to provide for the microbes as well as for the crop.

## *Contouring*

**Purpose of Contouring.** One effective way to reduce runoff and erosion is to place plant rows and tillage lines at right angles to the normal flow of surface runoff. The resistance to flow and the surface storage thus provided slow down the runoff and give the water more time to infiltrate into the soil instead of directly rushing off.

Since lines at right angles to the normal flow of runoff water are lines of equal elevation, or "contour" lines, this method of farming is called "contour cultivation," or "contour farming," or just "contouring."

While contouring is effective in conserving both water and soil, its main purpose varies with the climate, soil, and land use. On a pasture in a semiarid climate water conservation is the chief goal; in a cultivated field in a humid climate with many heavy showers soil conservation is the almost exclusive purpose of contouring. By and large, however, contouring serves a double duty, aiming at the conservation of both moisture and soil.

**Fig. 5-6.** Contour cultivation helps. Contouring in combination with basin listing on this Nebraska field has held most of the water of a 4-in. rain on a 7 per cent slope. (*USDA*)

**Effects of Contouring.** The primary effect of contouring is to create a large volume of detention storage on the soil surface. This results in reducing and slowing down runoff and consequently in a reduction of erosion. If the soil is permeable and its infiltration capacity high, a large part of the rainfall water enters the soil. On uncontoured land the same rain would cause runoff much sooner, and the soil particles contained in the water would fill up the pores of the soil surface, thus sealing it and lowering the infiltration capacity. This in turn causes still greater rates of runoff.

On slowly permeable soil the effect of contouring is much smaller because the surface storage is soon filled up, and since only little water enters the soil, much of it flows off superficially. This means that the effect of contouring on runoff is much greater on pervious soils than on those with relatively impervious subsoils.

In addition to providing surface storage, contouring causes runoff water to take indirect routes in its descent. This reduces the slope gradients the water follows and slows it down, thus spreading out the

**Fig. 5-7.** Change-over from straight-row farming to contour farming. The contour guidelines for strip cropping have just been plowed out on this hilly field in New York State. (*USDA*)

runoff over a longer period and decreasing its peak rate of runoff. This is particularly important in erosion control.

To be fully effective, contour-tillage operations have to be designed correctly. This includes both the direction of the operations and the capacity of depression storage that the tillage implement can produce. Once the storage capacity of contoured soil has been filled up and the infiltration rate has been reduced by long-continued rainfall, contouring loses its effectiveness. Further rainfall will run off as much as it would from uncontoured land. In fact, once the depressions are filled with water, the ridges may be washed away, and much of the water in these depressions is free to flow off and contribute to the total runoff.

**Place of Contouring.** Contouring fits best on cultivated sloping land. Land of use-capability class I is usually so level that cultivation can proceed in any direction without causing serious accumulations of runoff water, although it may be practical to cultivate in straight lines across the direction of the main slope.

Classes II and III represent the bulk of the land where contouring is needed and practical. As class IV land is generally kept in meadow or pasture, contouring is used only while it is in small grain in preparation for the new seeding. Class V land is normally not very steep and is covered by permanent vegetation on account of certain features that prohibit cultivation. This land neither needs contouring nor can contouring be practiced on it. If class VI is in pasture, it may benefit from contour furrowing in a dry climate, although shallow, stony soils may frequently make this difficult [4, 5].

Within the land-use classes II, III, and IV the desirability of contouring depends on soil type, land use, the configuration of the slopes, and the climate. Contouring is most effective on pervious soils of high infiltration capacity. Tight soils do not benefit much from contouring and may have to be protected by other measures. Where a large proportion of soil-conserving crops are grown, contouring is of lesser importance. Where the slopes are choppy and irregular, as on the so-called hummocky land, contouring is not practical because of the very many point rows resulting from the frequent changes in the direction of the contours. The more erosive the climate, the more important it becomes to use contouring.

**Contouring in Conjunction with Other Soil-conservation Measures.** Contouring together with ordinary good farming practices normally suffices to keep erosion on class II land down to a permissible rate. But even here sod waterways are frequently necessary to conduct the water safely off the field and to simplify the planting and cultivation of point rows. Where slopes are long, runoff water can accumulate and overtop the contour ridges. Under such conditions contouring should be supported by strip cropping, terracing, or basin listing. On class III land sod waterways and one of the other supporting practices have regularly to go with contouring. Where the erosion hazard on class IV land is particularly severe, strip cropping or mulch tillage has to be used in conjunction with contouring when the land is prepared for small grain. By and large the effectiveness of contouring is greatly reduced where the slopes are steeper than 10 per cent, because the volume of depression storage becomes too small.

**Laying Out Contours.** To lay out an individual contour line in the field, stakes are driven into the ground at intervals on points of equal elevation. These can readily be determined by the use of a level. Depending on the desired accuracy, an engineer's transit, a farm level, an Abney level, or a simple carpenter's level may be used. The latter has to be provided with two screw eyes for sighting and a small mirror for observing the position of the bubble.

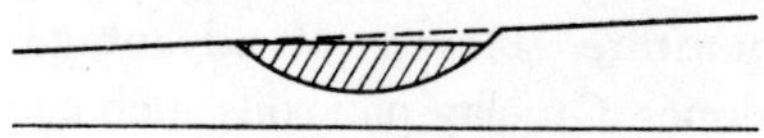

Depression on a 5 per cent slope. The largest part of the depression serves to hold water.

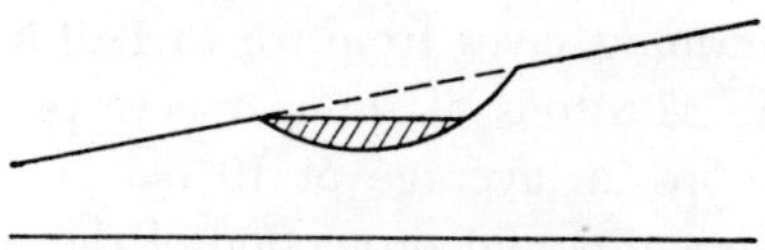

The same depression on a 20 per cent slope. Only a fraction of the depression holds water because much of its volume lies above the lower rim.

**Fig. 5-8.** The amount of depression storage decreases as slope increases. The same depression on a 5 per cent slope gives a much larger storage capacity for surface water than on a 20 per cent slope. The shallower the depression, the greater is this difference.

In designing a contour system several practical requirements have to be taken into consideration. If we adhere to the actual contours of the land too meticulously, farming becomes difficult because of the many point rows that result from the fact that true contour lines do not run parallel for any appreciable distance. It is necessary to compromise enough to get a reasonable number of parallel rows for seeding and cultivation before adjusting the neighboring lines to the contour. Generally, it is permissible to allow the rows to be as much as 3 per cent off the contour for short distances. In gently sloping fields it is usually sufficient to survey one or two contour lines and to cultivate parallel to these base lines and in straight lines on the flatter part of the land. Every field presents a different problem from that of any other field, and experience is needed to recognize the most practical solution.

Changing from straight-row farming to contouring generally involves a change of field boundaries and fences in order to avoid excessive turning and also because land-use-capability class boundaries frequently coincide approximately with contour lines.

Contouring means not only to plant on the contour but also to plow and cultivate on the contour. Rows of plants on the contour form an impediment to runoff, but if they are far apart, as corn plants usually are, this effect is quite small until cultivation has thrown up small ridges or terraces that actually create surface-storage capacity. This is the reason that land in contour-planted corn and land in straight-row corn usually have the same amount of erosion before the first cultivation, if other conditions are alike.

It is obvious that contouring results in an accumulation of soil moisture. This is of advantage on permeable soils on pronounced slopes. On clay-pan soils such accumulation may be harmful, especially on flatter land, sometimes resulting in yield decreases due to contouring in humid areas during periods of high rainfall.

Research in various parts of the United States shows that contour farming saves from 0.5 to 19.0 inches of water per year and from 0.3 to 41.8 tons of soil per acre per year [25, 58]. It increases yields of crops an average of 10 per cent as a consequence of moisture conservation and protection of the seedling plants from washing out, and it decreases the power requirement for tillage about the same percentage [53]. These figures, of course, vary greatly from case to case, depending upon rainfall, slope configuration, and other features.

**Disadvantages of Contouring.** Although contouring has qualities, this practice has disadvantages that reduce its usefulness. The difficulty of farming point rows has been emphasized before. The greatest hazard, however, is the possible accumulation of water in danger spots of the slope if the lines are off the contour for an appreciable distance. Adjustment of the tillage direction and establishment of sod waterways are the remedies. Sometimes terraces may be needed [15]. The accumulation of much water on a sloping field may be a dangerous condition in itself. If a heavy rain hits a contoured field just when it is soaked with water and all contour depressions are filled with water, a mass break of these many small reservoirs may occur, with catastrophic results. Building a curved contour fence requires special precautions. The sharper the curve, the less tension or stretch there should be in the wires and the closer the posts must be spaced. Otherwise, cold weather would contract the fence enough to pull over some of the more exposed posts.

## Contour Strip Cropping

**Purposes of Strip Cropping.** Contour strip cropping is the farming of sloping land in alternate strips of intertilled row crops and close-growing crops. These strips are placed at right angles to the direction of the natural flow of surface runoff water with the purpose of slowing down the runoff and of filtering out in the close-growing crop the

**Fig. 5-9.** Contour strip cropping. Wheat harvest in a contour-strip-cropped field in Oregon. (*USDA*)

soil washed from the land in the intertilled crop. In this way the rate of sheet erosion is reduced and gullying is practically eliminated. The same principle is used in two other soil-conservation measures:

*Field strip cropping*, the farming of relatively narrow fields of uniform width across the general slope but not closely following the contours.

*Buffer strip cropping*, the placing of sod-crop strips between contour-planted strips of the crops of a regular rotation. Frequently these buffer strips are located on steep or badly eroded areas that do not fit into the regular rotation.

Neither of these two methods is as effective as contour strip cropping, but they are helpful in spreading runoff water and therefore in reducing the tendency to gully formation. They also filter out some of the eroding soil.

Strip cropping slows down erosion, but allows it to continue. It provides no protection against raindrop splash on the intertilled strips.

Fig. 5-10. Soil conservation according to land-use capabilities. Contour strip cropping on land of capability class III, woods on class VII. (*USDA*)

Its greatest value lies in the fact that it decreases runoff rates, because the crop permitting much runoff is limited to half the hillside while a meadow crop that reduces runoff occupies the other half. Strip cropping is thus a very effective means of decreasing erosion hazards.

**Uses of Contour Strip Cropping.** *Crops.* Contour strip cropping is needed wherever intertilled crops must be grown on erodible land that cannot be protected by contouring alone [57].

*Land Class.* It fits best on land-use-capability class III and sometimes on class II. Also, the small-grain seeding on class IV land is best done by the strip-cropping method.

*Slope.* There are no definite slope limits for any soil-conservation measures, but by and large, strip cropping is used on slopes with 6 to 15 per cent grades or up to the limit of cultivation.

*Soil.* Strip cropping should be done on well-drained soil. It is expected to show at its best on soils which have low detachability and considerable amounts of abrasive-soil fractions of transportable sizes [22]. Such soil may gully badly if the abrasive particles are not removed by filter strips.

*Climate.* Contour strip cropping is of value wherever hard, beating rains are frequent and erosion is a hazard. In dry areas, however, winds often dessicate the many exposed borders of corn or other tall crops.

**Methods of Strip Cropping.** *Rotations.* The nature of strip cropping requires a strip of close-growing crops (meadow) to be grown side by side with a row crop. Consequently, a rotation used in a strip-cropping system has to contain at least as many years of meadow as of row crops. Otherwise, there are few limitations on the rotation used. In cases of a four-year rotation of corn, small grain, and two years of meadow ($C$-$g$-$M_1$-$M_2$), only two fields of approximately the same size are needed to allow each crop to occupy the same acreage each year. One field has strips of $C$ and $M_1$, while the other has strips of $g$ and $M_2$. This alternates each year. Each individual strip follows the desired rotation. Where a five-year rotation is chosen, five fields are needed to result in the same acreage for each crop, but for a six-year rotation, only three fields are required.

*Width of Strips.* This varies from 60 to 150 feet, depending on the rainfall, soil, slope conditions, and the crops grown. The more intense the expected rainfall, the more erodible the soil, and the steeper the slope, the narrower the strips should be. Since the strips vary in width owing to the irregularity of the slope, the farm size, and the width of the farming equipment, no specific recommendation can be made.

*Adherence to Contour.* The closer the strips are to the contour, the wider they can be. Irregularities of the slopes make it sometimes advisable to place corrective strips of sod between the row-crop strips. These correction strips permit the row crops to be planted practically on the contour. Observations of strip-cropped land have shown that locating strips deliberately about 3 per cent off the contour, with grass waterways at frequent intervals, may be of advantage. The implement marks serve as miniature terraces and carry away enough runoff to

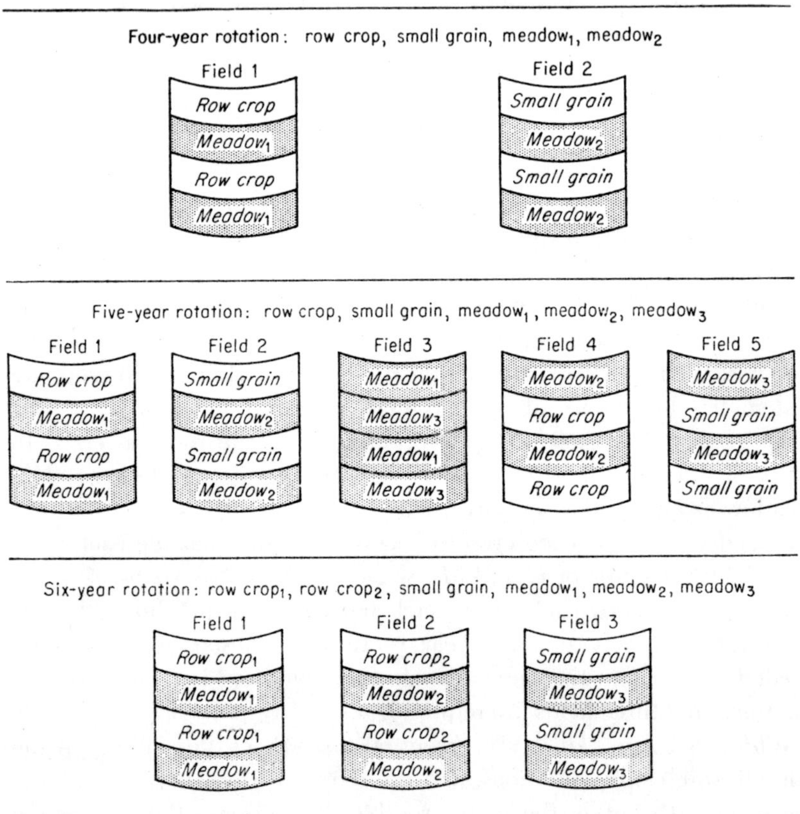


**Fig. 5-11.** Arrangement of rotations for strip cropping. In these three examples the land use is arranged so that the crops now on field 1 will be replaced next year by those on field 2, and so on, following the rotation.

prevent overtopping, but because of the small sizes of the channels, water flows slowly enough to avoid serious soil loss.

**Disadvantages of Strip Cropping.** If the meadow is to be pastured, the strips of row crop must be fenced out. This requires considerable lengths of movable fences. In general, farming operations are somewhat complicated by the need to move from one strip to the other.

Point rows, on the other hand, are easier to deal with on strip-crop land than on a solid-contour field, because turning can be done on the meadow strips.

Grasshoppers prefer field borders to the inside of corn fields; they cause more damage on strip-cropped corn than in a solid field of corn. In drier climates the exposure of much of the corn to the wind results frequently in losses from drought.

## *Terraces and Diversion Ditches*

**Purposes and Types of Terraces.** A terrace is an embankment of earth or stone constructed across the slope. The objective of terracing

**Fig. 5-12.** Conservation pattern. Terracing and strip cropping give this Texas land a handsome pattern. (*USDA*)

is to use sloping land more intensively than would otherwise be possible. This is done either by decreasing the slope over which the surface water runs or by intercepting the runoff before much of it can accumulate and carrying it to a safe outlet. In this way the rates of runoff and erosion are minimized.

There are three main types of terraces: bench terraces, irrigation terraces, and runoff-interception terraces.

*Bench Terraces.* The bench terrace is built by placing a wall of earth or stone along the contour and either filling in the space above this wall or allowing it to be filled by erosion [46]. The result is a

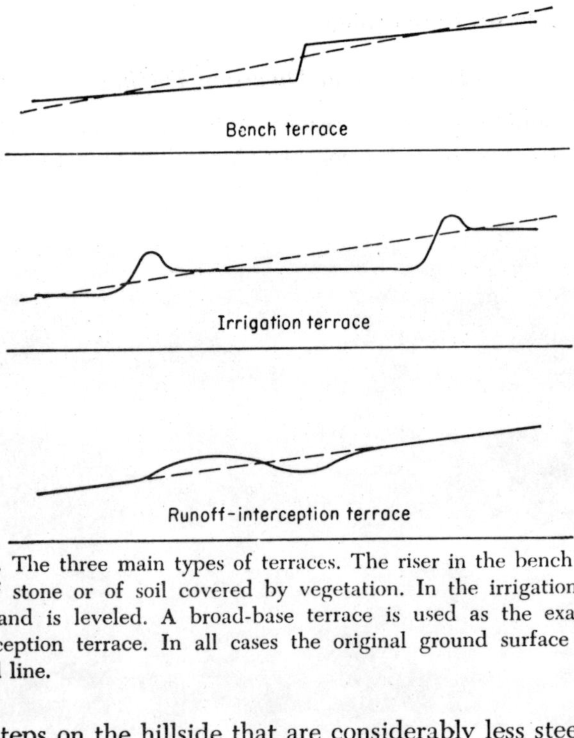


**Fig. 5-13.** The three main types of terraces. The riser in the bench terrace may be made of stone or of soil covered by vegetation. In the irrigation terrace the cultivated land is leveled. A broad-base terrace is used as the example of the runoff-interception terrace. In all cases the original ground surface is indicated by a dashed line.

series of steps on the hillside that are considerably less steep than the natural slope but allow runoff water to flow over the terrace wall and, consequently, down over the entire slope. This is the way most of the bench terraces in the Southeastern United States have developed. Erosion remains a serious problem, especially if the "risers" are not safely stabilized and runoff water is allowed to concentrate. Examples of this type of terrace can be found in the vineyards along the banks of the Rhine, in Italy, in the Southeastern United States, and on intensively farmed slopes in many countries. Bench terraces have a very long history. The Incas or pre-Incas of Peru built substantial terraces with stone faces, probably as long as four thousand years ago. Similar terraces in China, Japan, and the Philippine Islands date back more than two thousand years. However, they are no longer built in the United States.

*Irrigation Terraces.* The irrigation terrace is made by leveling out part of the slope. A contour wall is built on the downhill side of it,

**Fig. 5-14.** Bench terraces. Land in Southeast Asia is so valuable that the farmers build bench terraces to utilize even steep slopes for rice paddies. (*M. M. McClure*)

designed to retain water on the terraced land. It requires the greatest amount of work for construction and maintenance. By a series of flumes and weirs water is allowed to flow from the higher to the lower terraces for irrigation. When properly operated the irrigation-type terrace permits the growth of rice and other crops on steep mountain slopes with practically no erosion. It is in common use in Southeast Asia and other tropical areas. The irrigation terrace does not lend itself to the use of power tillage implements.

*Runoff-interception Terraces.* The runoff-interception terrace is an embankment or a ditch which intercepts the runoff from the land above it. Most interception terraces are designed so that excess water is carried off along the terrace channel. This is done specifically to prevent direct runoff of the rain water over the entire slope length. By preventing surface water from accumulating over a long slope and making it flow in a nearly level channel to an erosion-resistant outlet, the water is slowed down to such an extent that it causes very little erosion. Terraces, however, have no influence on detachment of soil

**Fig. 5-15.** Narrow-base terrace. Terracing is particularly effective in connection with contour cultivation. Narrow-base terraces cannot be planted and cultivated with large modern farm machinery. They are useful in one row ("one mule") operations. (*USDA*)

from rain-drop impact except when water concentrates behind them and forms a protective layer over part of the area.

Originally, interception terraces were built either as channel terraces or as narrow-base ridge terraces, leaving a ditch or channel to catch runoff water. The soil material of the channel terrace is thrown downhill. The main purpose of building channel terraces is to intercept the runoff water and to get it off the land without having it run over the full length of the slope. The channel terrace is therefore also called the "drainage-type terrace."

The narrow-base ridge terrace is constructed with its ridge well above the original ground surface, with as little channel as possible. Its main function is to impound water over a wide area above the ridge, so that a maximum of water is absorbed.

Both the channel terrace and the narrow-base ridge terrace are no longer extensively used in the United States, largely because farm

**Fig. 5-16.** Broad-base terrace. Terraces are used in combination with sod crops in the rotation to protect this rolling farm land in Wisconsin. (*USDA*)

machinery will not operate on the narrow, steep ridge or deep channel. To overcome this objection a broad-base terrace has been developed. It is a wide ridge with gentle side slopes. Even large modern implements can function properly on it. Its original design is credited to P. H. Mangum, a North Carolina farmer who first used it in 1855. The present-day "terrace" that is universally used in the United States is a modification of the Mangum terrace. The first research work on drainage terraces to serve as a sound scientific basis for the practice was started by C. E. Ramser [46] in 1915. Minor improvements in design, operation, and maintenance of interception terraces are being made as their use increases.

Generally, interception-type terrace channels have a slight grade to carry off the water. Level terraces are used in dry regions on permeable soils primarily to impound water and to make it all enter the soil. In the experiments at Spur, Texas, a field with level terraces had no runoff and no erosion beyond its limits for a period of twenty years [19]. Basin listing, or the construction of depressions on the contour, is done for the same purpose as level terracing.

Diversion ditches are similar to channel-type terraces. Generally, they are designed to carry more water. Their purpose is to divert runoff from an area above a cultivated field. Since the broad-base terrace is the predominant type of terrace used in the United States, most of the following discussion refers to it.

**Use of Terraces.** Terraces should be put where slopes are so long that an excessive amount of surface runoff would accumulate and cause erosion. This frequently occurs on land-use classes II and III on slopes between 2 and 8 per cent grade. Where slopes are very long, terracing is done on land of only ½ per cent grade. Where there is a dearth of flat cropland, slopes up to 10 or even 16 per cent grade are terraced. On slopes of 12 to 16 per cent, it is difficult to build and maintain terraces that have adequate capacity and can be farmed with modern machinery. These slopes are too steep for cultivated crops and should not be used for such unless conditions require it [29].

The shallower the soil profile, the flatter the land has to be for terracing. On steep, shallow soils too much of the underlying strata would be exposed. Soil that has a tendency to slip should not be terraced. This is the case where bedrock or a tight clay layer are near the surface. Since terracing controls transportation of detached soil particles but not detachment by rain splash, it is especially effective in controlling erosion on soils of low detachability and high transportability. Graded-channel terraces belong in a humid climate. Absorption-type terraces (ridge terraces) with small grade or completely level belong in a semiarid climate. Bench terraces belong in a densely populated country, because they make it possible to utilize very steep land for agriculture.

**Effectiveness of Terraces.** *Erosion Control.* The amount of erosion control provided by terracing depends on many factors. By and large, terracing reduces erosion on cultivated land to one-fifth or even one-tenth of what it would be if no control measures were used. Consequently, terracing permits a much more intensive land use [54]. The steeper the slope, however, the less efficient terracing becomes.

*Water Conservation.* By slowing down the velocity of runoff, terraces give water more time to soak into the ground. On the other hand, on a wet soil water is carried off sideways in the channels and does not spread over the entire field; thus more water is carried off directly into the outlets. In this way terraces serve to maintain a desirable intermediate moisture content in the soil.

**Disadvantages of Terraces.** If not properly constructed, terraces may cause water to accumulate in the low spots of the channels and to spill over the ridges and wash them out. Thus water may be dumped from terrace to terrace, causing severe gullying.

In building terraces, infertile subsoil is frequently exposed in the channels. Therefore crop yields in the channels are low during the first years. Later on, sediment improves the soil conditions.

Obviously, terrace construction costs money, and operation of a terraced field requires special attention. All these disadvantages are not serious enough to overbalance the great advantages of terracing, where it is suitable.

**Design of Broad-base Terrace.** A terrace must be designed and built so that its channel can carry all the water that can be expected to reach it without the danger of overtopping. For this purpose it has to have an ample cross-sectional area and sufficient grade to keep the water moving toward the outlet without eroding the channel. In order to avoid excessive sedimentation in the channel, its size must be nearly uniform throughout its length. Since the amount of water that has to pass through the channel increases as it approaches the outlet, the channel grade is frequently made steeper toward the outlet, especially in longer terraces. Channel grades vary from 0.1 to 0.4 per cent. The more absorptive the soil, the more nearly level the grade is. As previously mentioned, in semiarid land terrace channels may be horizontal, and the ends closed to store all the water.

The cross-sectional area is frequently around 12 square feet. It varies from 8 to 16 sq ft, depending upon storm characteristics of the locality, erodibility of the soil, and the type of land use. To make larger terraces would require too much earth movement and would expose too much subsoil. Instead of making terraces larger, they are

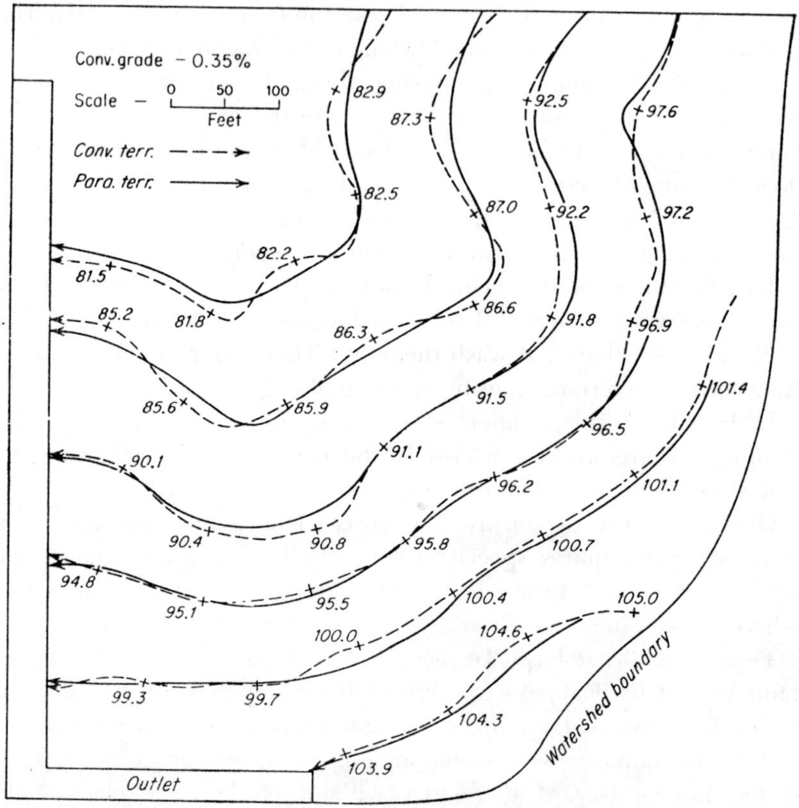


**Fig. 5-17.** Parallelized-terrace-system study. The three maps give an example of using conventional and parallelized terraces on a field. The area in point rows is greatly reduced by the latter method. (*R. P. Beasley and L. D. Meyer: A new terrace construction technique, Agr. Eng.*, **38**:32–36, 1957)

**Fig. 5-17*a*.** Map of conventional terrace system including elevations along the terraces, and parallelized terrace system designed from this map.

merely placed closer together, so that they have to carry less water.

The total width of a terrace is from 15 to 40 ft. The side slopes should not exceed 20 to 25 per cent.

A terrace should be no longer than 1800 ft, preferably no longer than 1200 ft [37]. The higher the infiltration capacity of the soil, the longer the terrace can be. To avoid extreme lengths it is possible to let the terrace grade toward two outlets. Terraces that are longer than 400 ft require increasing channel grade toward the outlet.

**Designing a Terrace System.** The steeper the land, the closer together terraces must be to be effective. Again, the more intensive the land use and the less pervious the soil, the closer together the terraces should be.

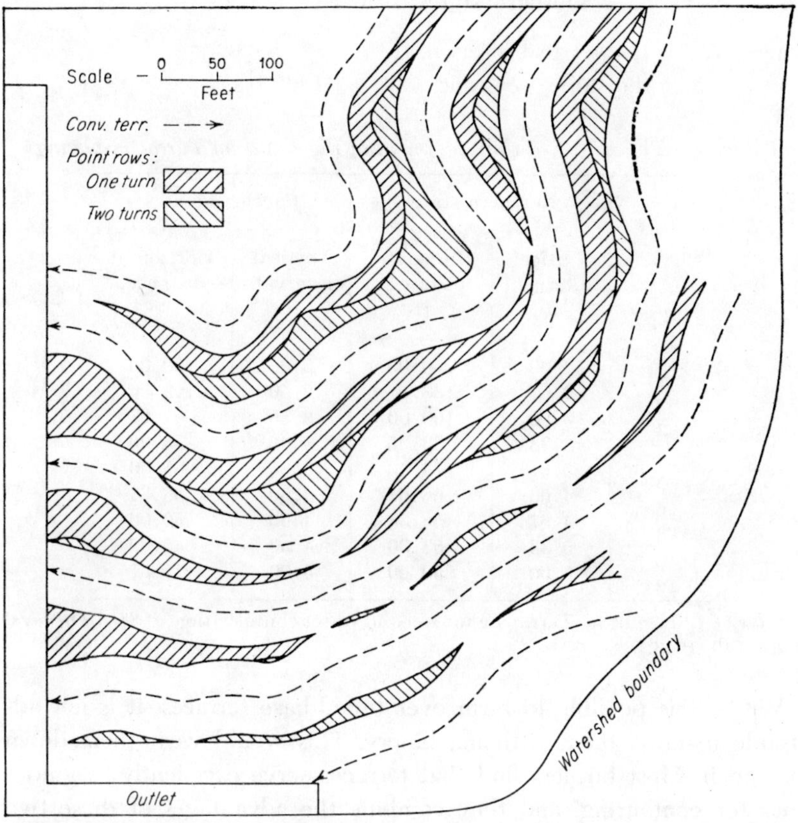


**Fig. 5-17*b*.** Conventional. Point-row area of conventional terrace system shown in Fig. 5-17*a*.

There is a lower limit for terrace spacing. When land with over 12 per cent slope is terraced according to standard design, too much land surface is disturbed.

The best spacing will consequently vary from case to case. Since the amount and intensity of rainfall have a very pronounced effect on the erosion hazard, they have to be taken into account in arriving at the proper spacing of terraces. The U.S. Department of Agriculture

[29] has therefore suggested the following spacing, depending upon whether the terraces are to be built in the Northern or in the Southern states:

$$\text{Northern states: V.I.} = 2 + S/3$$
$$\text{Southern states: V.I.} = 2 + S/4$$

where V.I. = the vertical interval, ft
S = the slope, per cent

*Table 5-2. The Effect of the Steepness of the Land on Terrace Spacing*

| Slope, % | Northern states | | Southern states | |
|---|---|---|---|---|
| | Vertical interval, ft | Horizontal distance, ft | Vertical interval, ft | Horizontal distance, ft |
| 1 | 2.00 | 200.00 | 2.00 | 200.00 |
| 2 | 2.67 | 133.5 | 2.50 | 125.00 |
| 3 | 3.00 | 100.00 | 2.75 | 91.67 |
| 4 | 3.33 | 83.25 | 3.00 | 75.00 |
| 5 | 3.67 | 73.40 | 3.25 | 65.00 |
| 6 | 4.00 | 66.67 | 3.50 | 58.33 |
| 8 | 4.67 | 58.38 | 4.00 | 50.00 |
| 10 | 5.33 | 53.30 | 4.50 | 45.00 |
| 12 | 6.00 | 50.00 | 5.00 | 41.67 |

After C. L. Hamilton, Terracing for soil and water conservation, *USDA Farmers' Bull.* 1749, 1943.

While it is possible to farm over broad-base terraces, it is not advisable to cross them with machinery. This would wear them down too much. Most farmers find that terraces serve excellently as guidelines for contouring, and they combine the advantages of these two soil-conservation methods. If terraces follow the true contours—except for the slight grade required—the spaces between them vary greatly in width, except in the few cases where slope gradients are fairly uniform. This results in many point rows which are very inconvenient to farm. A "parallelized terrace system" has been devised that reduces the number of point rows required [2]. First, the terrace lines are laid out in the conventional manner and each is smoothed to eliminate sharp curves and minor irregularities. Next, two men walk adjacent terraces, holding a stretched tape between them.

The terrace lines are adjusted to be equidistant wherever possible, unless the deviation from the contour becomes too great.

In this way terraces are parallel for some distance, and therefore less point rows are needed. Since this system causes terrace lines to

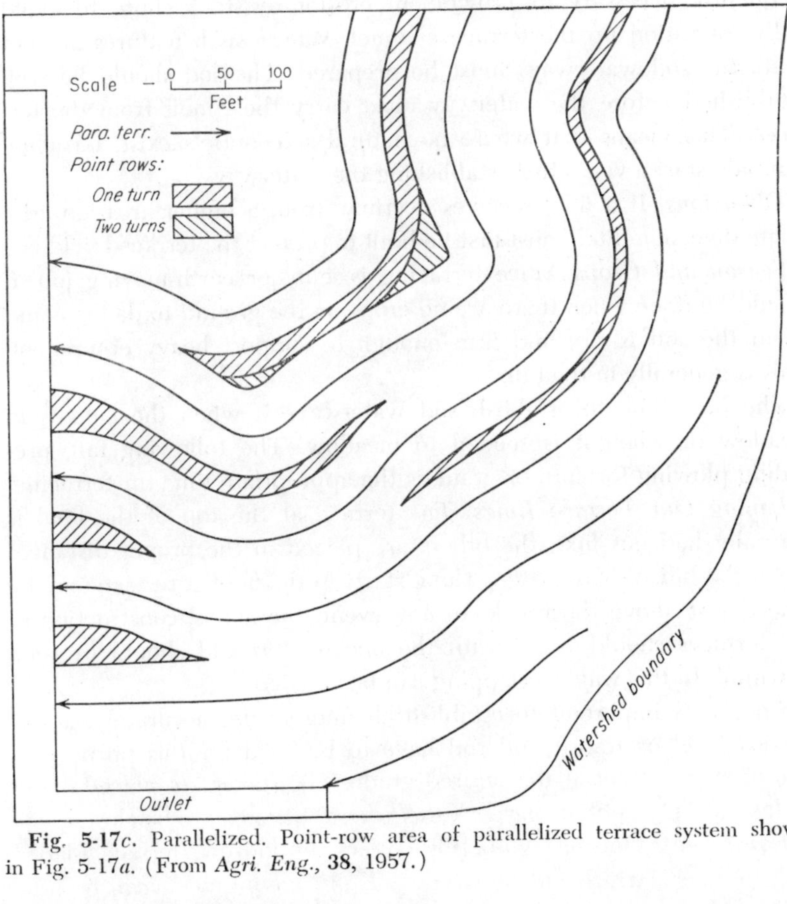


**Fig. 5-17*c*.** Parallelized. Point-row area of parallelized terrace system shown in Fig. 5-17*a*. (From *Agri. Eng.*, **38**, 1957.)

deviate from the contour, more earth movement is required to give the terrace channels the specified grade. Both the laying out and the construction of parallelized terraces are more costly than the conventional method, but the farming is greatly simplified because of the smaller number of point rows. In a large field the entire area frequently does not need to be terraced, as some of the land may be flat.

**Building and Maintaining Terraces.** *Outlets.* Before any construction work is begun, outlets for the safe removal of the water must exist [30]. In some cases the water can be conducted directly into a natural draw or a permanent pasture or woods. Occasionally, road ditches, gravel pits, or limestone sinkholes can be used. However, it is frequently necessary to provide an erosion-resistant chute to avoid gully formation up the terrace channel. Where such features are not available, sod waterways must be prepared. The sod should be well established before the waterway must carry the runoff from the terraces. This means that where no natural safe outlets exist, terracing can only start a year after establishing the waterways.

*Diversions.* If a field receives overflow from a higher area, an adequate diversion ditch must first be built to protect the terraced field.

*Season and Crops.* Since terracing is a major earth-moving job, it should be done when there are no crops on the ground to damage and when the soil is dry and firm enough to support heavy equipment. This is generally in the fall.

The best time to establish sod waterways is when the land is in meadow or when it is seeded to meadow. The following fall, preceding plowing for corn or grain, is the appropriate time for terracing.

*Laying Out Terrace Lines.* The terrace at the top of the field is normally laid out first; the others are placed in the proper distances down the hill. Where a steep slope starts in the field, a terrace may be placed just above the break. In any event, the actual construction of the terraces should begin with the one on top and should proceed downhill. In this way overlapping can be avoided.

Since it is important to establish channel grades accurately, a surveyor's level or transit and rod have to be used for this purpose. A line of stakes is set at the desired grade. The stakes are placed about 50 feet apart, except at sharper turns where they need to be closer.

*Equipment.* One or two-bottom plows, bulldozers, terrace blades, disk terracers, whirlwind terracers, and other equipment can be used to build terraces. The selection depends on the availability of the equipment and whether the farmer prefers to do the work himself or wants to have it done by specialists. Where gullies are a problem, bulldozers can be used to make the necessary fills as well as to build the terrace. On more gentle slopes the moldboard plow is very satisfactory. Where much terracing is to be done, specialized equipment is preferable.

*Construction.* The methods of building terraces depend on the equipment used. The so-called "island method" is used with the moldboard plow on slopes up to 6 per cent. Going around the "island" of

**Fig. 5-18.** Whirlwind terracer. A vertical rotor behind a single bottom plow can be operated at three speeds. The whirlwind terracer gives a quickly settled terrace because the soil particles are finely broken. (*USDA*)

unplowed land the plow throws the earth toward the center. This operation is repeated until ridge and channel are formed. Twenty to thirty rounds may be necessary. The more specialized equipment throws the earth farther and moves more earth. Therefore less rounds are needed. The ridge has to be made higher than designed to allow for settling.

*Maintenance.* To assure proper functioning of the terrace system, frequent attention has to be given to the maintenance of terraces and outlets. This is particularly important after heavy rains in the first year after construction. Wherever there is settling or washing of the ridges, they must be raised with new earth. The main danger points are the places that received fills.

The general maintenance of terraces consists in plowing toward the ridge in order to maintain the proper height.

**Costs of Terracing.** As conditions of soil, erosion, slope, and stoniness vary greatly from field to field, so does the cost of terracing. J. H. Yeager et al. [61] found that the average cost of terrace construction

**Fig. 5-19.** Terracing with a bulldozer. Bulldozers are effective in terrace construction, especially where gullies or uneven slopes require much earth movement. The cost per mile of terrace is high. (*USDA*)

in western Indiana, including the necessary preparation of the field, cost $3.71 per 100 linear feet. The cost is on a 1949 basis. This cost includes 2 man-hours of work.

The cost of constructing terraces is substantially lower in a field with smooth surface. The bulldozer work needed to fill in gullies represents a large part of the expense.

Cost of maintenance, especially after the first year, is only nominal if sound farming practices are followed in connection with terracing.

Compared to the advantages of terracing—where it is suitable—its costs are small.

### *Sod Waterways*

**Purpose and Value of Sod Waterways.** Sod waterways are smooth channels lined with sod for the safe removal of surface runoff water from cultivated land. Grass sod is the most satisfactory and economical protection for such waterways.

Sod waterways are established wherever the danger of gullying exists. They are needed where surface runoff has the tendency to accumulate in a cultivated field. They serve excellently as supplements to contouring and strip cropping and are indispensable as terrace outlets where no natural water disposal exists. But sod waterways are useful without reference to any other soil-conservation measure.

**Fig. 5-20.** Sod waterway. The runoff from this Maine potato field is led off harmlessly in a sod waterway. (*USDA*)

By protecting the soil in the waterways that are normally most exposed to erosion, the slope gradients near the waterways are maintained, and consequently the erosion hazard is not increased. Sod waterways are used on land-use-capability classes II, III, and IV. They are placed either in natural draws or along field boundaries, especially if they are to serve to remove runoff water from terraces. It is sometimes wise to adjust fields so that the waterways form the boundaries. Sod waterways fit best where runoff is intermittent, since permanent streams kill the grass.

**Design.** A sod waterway has to be designed so that it can carry all the surface runoff safely without eroding. It must be deep and wide enough both to carry the runoff and to be easily crossed with tillage implements. The first thing to consider is the expected maximum flow.

Fig. 5-21. Double-duty waterway. This sod waterway serves as a drainage ditch as well as to control erosion. (*USDA*)

This depends on the size and steepness of the drainage area, the absorptive capacity of the soil, the surface storage, the land use and treatment, the presence of other soil-conservation measures, and the expected rainfall. Generally, waterways are designed so that they can carry the peak flow that can be expected in ten years [47]. To design them for runoff rates that might happen once in fifty or even in one hundred years would require uneconomically large waterways.

The majority of sod waterways have either parabolic or trapezoidal (flat-bottom) cross sections. The parabolic ones are used in natural draws, while the trapezoidal cross sections are preferred on smooth areas, such as those used for terrace outlets at the edge of a field. In a waterway with a flat bottom the danger exists that a slight impediment to flow may cause the water to meander and to cause erosion.

The desirable velocity of the flow in the channel should be low enough not to cause erosion, yet high enough not to cause sedimentation. This means that under most circumstances it could vary from 2 to 3 feet per second as the lower limit to 5 to 6 ft per sec as the higher limit. It is desirable to maintain a uniform velocity throughout the length of the waterway. The size and shape of the cross section help to regulate the velocity. The channel slope can vary between 0.5 and 10 per cent; in extreme cases it can go up to 14 per cent. Where the slope is steeper, it is necessary to build erosion-control dams or chutes to decrease the gradient of the waterway. There is no definite maximum limit on the length of a sod waterway, but it usually is not practical to extend them beyond half a mile, because they would have to be very wide to be safe. It is better to put the entire lower stretch into permanent pasture or meadow. A short waterway can have the same cross-sectional area throughout its length. In a long waterway this area has to increase downhill with the amount of water it must carry. Sharp bends have to be avoided, as these might cause erosion of the outside bank. It is generally preferable to give waterways a smooth course. Making them entirely straight, however, might cause excessive flow velocities.

**Construction.** Ordinary farm tools, such as plow and slip scraper, can serve to construct waterways, but it is preferable to use a grader blade or a bulldozer, especially if considerable earth movement is needed. Since the channel is particularly subject to erosion before vegetation has been established, it is well to construct it when the field is in meadow, and the amount of expected runoff is the least. If the erosion hazard is great, the runoff should be diverted from the waterway until a good sod exists.

**Vegetation.** *Species.* As the name "sod waterway" implies, the best vegetation to protect waterways from erosion consists of sod-forming grasses. The main requirement is that the species is adapted to the climate and to the site [28].

Examples of grasses that have done exceptionally well are Kentucky bluegrass (*Poa pratensis*), smooth brome grass (*Bromus inermis*) and meadow fescue (*Festuca elatior*) in the Northern states, western wheat grass (*Agropyron smithii*) and sand bluestem (*Andropogon hallii*) in the drier portions of the Great Plains, Bermuda grass (*Cynodon dactylon*) in the South, and reed canary grass (*Phalaris arundi-*

*nacea*) in wet locations. Other grasses that have been used successfully in waterways are timothy (*Phleum pratense*), red top (*Agrostis alba*), common rye grass (*Lolium virgatum*), and intermediate wheat grass (*Agropyron intermedium*). Generally, the rhizomatous grasses are preferred because they spread underground and provide safer protection than the bunch grasses. Deep-rooted legumes are seldom used in sod waterways. They have the tendency to loosen the soil and to make it more erodible under the impact of fast-flowing runoff water. Occasionally, white clover is seeded along with grasses, but it is of little value to improve the efficiency of the sod. A light seeding of small grain is sometimes used to provide a quick cover before the grasses are established.

*Establishment.* As the erosion hazard in waterways is great, everything has to be done to establish quickly a complete sod cover of the ground. A moist, fertile seedbed is necessary. The ground should be heavily fertilized and thoroughly worked. It is best to disk about 10 tons of manure per acre into the soil, leaving some of it on the surface as a protection for the grass seedlings. A light coat of manure or straw mulch is sometimes used after seeding. Where the danger of blowing or washing is great, oil emulsions or asphalt are used to keep the mulch in place. The grass is seeded at about twice the rate of meadow seedings and is lightly covered with a cultipacker. It is generally best to seed more than one species, since the site conditions are not uniform throughout the waterway and the weather cannot be predicted.

The late summer or early fall is the preferred time to shape waterways, because the ground is dry. Grass seedings are best done in fall. This is the time when grass is established most easily and when no excessive rainstorms are anticipated that might endanger the waterway. In cases of particularly great erosion hazard, it may be necessary to resort to sodding instead of seeding the waterway. Because of the great expense involved, usually only partial sodding is done, and the rest of the area is seeded.

**Maintenance.** The grass has to be kept short and flexible, so that it shingles as water flows over it but does not lodge permanently. It should be mowed two or three times a year. The hay must be removed, so that it does not pile up at some spots in the waterway and form an obstruction to flow. This would reduce the capacity of the waterway and might divert the direction of the water and cause turbulence and cutting of the channel. It is also possible to keep the grass short by light pasturing. This has to be done carefully; never in wet weather and never by hogs. When the field is pastured, it may be necessary to apply manure on the waterways to discourage excessive grazing. Waterways should not be used as runways for livestock or as roads. After the sod is once established, only light fertilization is needed, because the runoff brings plant food from the watershed above.

When crossing waterways, tillage implements should be disengaged, plows lifted, and disks straightened. Tillage operations should be on the contour, never up and down the sides of the waterway. It is essential that any needed repair is done quickly, so that the damage does not become too large.

### *Gully-erosion Control*

Gullies are formed where runoff water concentrates on insufficiently protected land. Most active gullies are therefore found in cultivated fields and in overgrazed pastures. They seldom occur in well-managed pastures or in forest land. Gullies form on any soil, but particularly on soils that are slowly permeable and of high detachability. Where both surface soil and subsoil are easily eroded, gullies with U-shaped cross sections develop. Where the subsoils are resistant to erosion gully, cross sections are usually V shaped.

In order to prevent gullying or to heal a gully, two main approaches are used [35]: (1) reducing the energy of runoff water and (2) protecting the waterway.

To reduce the energy of runoff water, its amount or its velocity must be reduced. This can be accomplished by better land use and treatment or by diverting the water by means of terraces before it enters the drainage way. It can also be done by widening the waterway, so that the water will flow shallower.

The waterway may be protected by vegetation alone or a combination of vegetation and structures.

Any method that helps to increase the infiltration capacity of the soil is helpful in the control of gullying. Probably the surest method is the extensive use of erosion-resistant vegetation in the drainage area above the danger spot. Where row crops or other erosive crops are used above a potential gully, soil- and water-conservation methods, such as contouring, strip cropping, mulching, and terracing, should be used to minimize the amount and rate of runoff.

To protect a waterway from gullying, three main approaches can be used [34]. The waterway can be shaped to widen the channel, so that the water flows slower and at a shallower depth and the slopes

**Fig. 5-22.** Dry masonry dam. This rock dam was built in 1936 by members of the Civilian Conservation Corps in Oklahoma. Photographed twenty years later, it shows that it has served well to retain soil in the gully and to permit a dense stand of grass to be established. (*USDA*)

of the gully are less steep. The second means is to vegetate the waterway as discussed previously. Where these two methods are not enough, steplike structures have to be built to take the runoff water down safely. The main purpose of gully-control structures is to reduce the slope of the drainage channel. This makes it possible to establish and maintain vegetation that will protect the waterway from erosion. A variety of designs for such structures exist. The main features of all of them are that they cause all the runoff water to flow through the structure, take it safely to a lower level, and let it dissipate most of its kinetic energy within the confines of the structure, so that it will cause no erosion immediately below the structure. If the main objective of building the gully structure is not reached and the slope of the channel remains too steep, the danger exists that the gully will cut back and undermine the structure. In some cases such structures can be of a temporary nature. They may be built of wood, brush, or wire. Temporary structures are only of value where there are temporary erosion

**Fig. 5-23.** Vertical-drop spillway. The concrete-weir notch dam drops the water safely to the apron on the lower level and provides all excellent tile outlets. (*USDA*)

problems. They fit where improved land use and improved conservation methods are planned. After a temporary structure has disintegrated, the slope of the waterway is as steep as it was before, and the gully problem is as serious. Where no pronounced soil-conservation measures are anticipated that can protect the waterway by themselves, permanent structures are preferable. Since gully-control structures interfere with tillage and other farming operations, it is best to place them at field boundaries.

To prevent the formation of gullies a variety of concrete and masonry structures can be used. Vertical-drop spillways, also called "notch-spillway dams," are used where the drainage area is relatively large and the head is not higher than 10 feet. In case of high heads, slanting chutes are used for large drainage flows and drop-inlet pipe-conduit structures for smaller areas. Box-inlet drop-spillway structures are preferred where it is necessary to concentrate the issuing water from a large watershed into a relatively narrow channel. A sod chute

**Fig. 5-24.** Reinforced concrete chute. A slanting flume lowers the water from a terrace outlet through an 8-ft fall. (*Portland Cement Association*)

must be protected by a toe wall, especially if the water runs through it for long periods during the year.

In all cases a stilling basin or an apron serves as energy dissipators. The water issuing from the structure is always directed parallel to the direction of the waterway in order to avoid cutting of the sides. Rather efficient gully structures can be built of living stakes of willows, cottonwood, or similar species. As these stakes take root and grow, the structures become safer. This type of protection, however, has only a limited application, fitting best into reforestation areas, wildlife refuges, and pastures. In this latter case the structure has to be fenced from grazing animals.

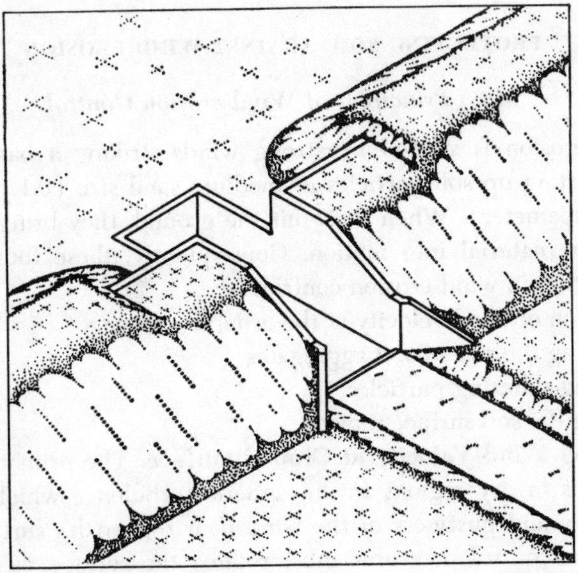

**Fig. 5-25.** Box-inlet-drop spillway. Water from a wide approach is narrowed into a channel as it is dropped. (*USDA*)

**Fig. 5.26.** Toe wall. A toe wall helps to keep the sod chute in place, preventing undercutting. (*Portland Cement Association*)

## PROTECTING SOIL AGAINST WIND EROSION

### *Main Principles of Wind-erosion Control*

Wind erosion is started by strong winds striking a bare, dry soil surface, lifting up soil particles of medium sand size (0.1 to 0.5 millimeter in diameter). When these hit the ground, they bring both finer and larger material into motion. Consequently, these four measures are important in wind-erosion control:

Reduction of wind velocity at the ground surface
Increasing the size of soil aggregates
Trapping saltating particles
Keeping the soil surface moist

**Reducing Wind Velocity at Ground Surface.** The primary cause of dust storms in dry regions is atmospheric turbulence which increases the velocity and gustiness of the wind near the earth's surface. Effective erosion-control measures must reduce the surface velocity sufficiently to reduce or eliminate initial movement of soil particles in saltation.

Cultural practices which tend to reduce turbulence are effective for soil-drifting control. However, some of the successful methods now in use are based on the principle of placing obstructions in the path of the wind to break its force. Such obstructions may increase the turbulence, but they absorb much of the force of the wind, thus reducing its velocity.

Convectional eddies increase the surface velocity of wind and may cause considerable erosion, especially on bare or fallow soil. Soil without vegetative cover absorbs the sun's rays and is frequently warmer than the air above. Vegetation absorbs the sun's rays, and the potential temperature difference between the air and soil is not so great. In strip farming, convectional eddies produced over the heated fallow surface are damped down somewhat as they pass over the adjacent cooler strips of vegetated soil.

The velocity of wind increases rapidly with height above the surface of the ground. Wind velocity is higher over high ground, such as knolls, than in lower positions. Therefore knolls tend to drift first [20]. Knolls should be stabilized with permanent vegetation where possible.

**Increasing Size of Soil Aggregates.** It has been stated previously that the most erodible soil is composed of fine granules varying in size from 0.1 to 0.5 millimeters in diameter. One approach to wind-erosion control is the creation of larger aggregates which are less susceptible to movement by wind.

The decomposition of organic matter may or may not have an effect on the resistance to wind erosion, the deciding factor being whether the granules formed are sufficiently large to withstand the erosive force of wind. In many of the prairie regions of North America, humus formation results in the development of fine granules which are susceptible to wind erosion. However, some of the clay soils which are also high in organic matter are coarsely granulated and are not normally subject to serious wind erosion. In areas where soil drifting is a problem, the addition of small amounts of calcium carbonate causes fine-textured soils to form small granules and thereby become more erodible.

Newly broken sod, with its high content of fibrous roots, is seldom subject to severe wind erosion. In time, however, the roots decompose, and the soil becomes erodible. Cultivation expedites the decomposition of organic matter and reduces the size of soil aggregates and thereby contributes materially to soil conditions which are susceptible to wind erosion. This is also true of some fallow land. Cultivation is necessary for the control of weeds, but excessive cultivation will increase soil drifting.

The repeated action of frost causes a breakdown of the larger aggregates and simultaneously causes fine particles to stick together in saltation-size fine granules; therefore frost is a factor contributing to increased susceptibility to erosion. Soil cover, such as sod or a mulch of crop residues on the soil surface, can be very effective in preventing the periodic formation of ice crystals in the soil which tend to break up the large soil aggregates [13, 14].

When pulverized soil is wetted and dried, fine particles adhere and form a crust which is resistant to movement by wind. Repeated wetting and drying, and especially freezing and thawing, break up the crusts and clods and increase erodibility.

**Trapping Saltating Soil Particles.** The fine dust particles and the coarser granules and particles are not normally moved by the direct action of the wind. Rather they are started in motion by the particles in saltation. The obvious conclusion is that if we want to stop wind erosion, we must first stop saltation.

Saltation can be effectively eliminated by a vegetative cover or thick stubble. However, on dry, bare, or fallow soil, saltation cannot

**Fig. 5-27.** Trapping saltating particles. Deep listing traps some eroding soil particles. Wind erosion is not completely stopped by such practices. (*USDA*)

be completely eliminated. The action of saltating particles on other particles can be partially controlled by catching them before they become suspended or by trapping them as they strike the ground. Strips of upright-growing crops or stubble across the direction of prevailing wind have proved successful in catching the abrasive material carried by saltation [48].

Another method of removing abrasive materials is to trap it in terraces, furrows, ridges, lister basins, etc. [23]. Where trap strips are used, the required width of the strips depends partly on the length of jumps of grains in saltation and partly on the trapping capacity or receptiveness of the surface. Standing grain, with its stubble, is probably the most effective form of trap, for it will trap all of the soil moving into it [9]. The minimum width of the strips of cultivated and uncultivated land must be the maximum width that a particle can move in a single leap of saltation plus a reasonable safety factor.

To jump a 10-foot strip of short stubble, a grain would have to rise vertically at least 12 inches. Since 90 to 98 per cent of the flow in saltation is below that height, a 10-ft trap strip is usually at least 90 per cent effective in controlling saltation. With very high-velocity winds the strip width would have to be increased. For example, to be

**Fig. 5-28.** Wind-erosion control. A 30-mph wind did not erode this field which was protected by sorghum stubble. (*USDA*)

totally receptive a 20-ft strip is needed for a 25-mile-per-hour wind [10].

The height of the vegetation or stubble on the trap strips also affects the effectiveness of the strip. The holding capacity of a strip varies with the height of the cover on the strip.

**Keeping Soil Surface Moist.** As long as surface tension holds the soil grains together, wind cannot lift them. Water is needed for surface tension to exist in the soil. Any method that will maintain the soil surface in a moist condition helps to prevent wind erosion. Irrigation and crop-residue mulch are probably the most commonly used techniques of keeping the soil surface moist in areas that are subject to wind erosion. The increase of surface tension between the soil grains is, of course, only a by-product in either case. The prime purpose of both is to provide a water supply for the crops. In the case of mulch it is also aimed at breaking the force of the wind at the ground surface. By and large, it is impossible to keep cultivated, exposed soil moist in a dry climate.

## GENERAL METHODS OF GOOD AGRICULTURE

### *Land Use*

Strong winds cannot be stopped, but drifting can be controlled by good management and proper handling of crops and soil. Control

**Fig. 5-29.** Wind erosion as a result of successive wheat failures. Uncontrolled erosion has made living impossible. The house has been vacated for two years. There is no vegetation to stop the hundreds of sand drifts. (*USDA*)

measures are simple and need not upset the farmer's normal program. The first approach to the problem is wise land use. In some of the more severely drifting areas cooperation between farmers is essential, because soil drifting on one farm may interfere with control measures on an adjoining farm.

In part of the Great Plains the relatively low average annual precipitation and long dry periods in the spring and early summer make the problem very difficult. Soils drift more readily when dry. Such dry periods often result in serious damage to newly seeded crops, even on medium- and fine-textured soils. There are some sandy areas that drift so readily that it is more profitable to keep them in meadow

**Fig. 5-30.** Wind erosion is controlled with a cover crop. This is the same area as shown in Fig. 5-29. Hummocks have been leveled and planted to milo. Judicious cropping will help to maintain the body and the productivity of this field. (*USDA*)

or other permanent vegetation than to cultivate them. On many farms there are areas that are not suitable for cropland. Soil may be classified as to its capabilities and limitations and handled accordingly [49].

Land not suited for frequent cultivation can be used for pasture, hay, or woods. The farmer will be influenced in his choice by economic and ecologic considerations. On the more erodible cropland, rotation with a maximum proportion of hay or pasture should be practiced.

The land less subject to drifting may be used for more intensive rotations with intertilled crops. The number of years in hay crop will vary with erodibility of the soil. In some cases it may be only two years out of five; at the other extreme it may be advisable to use several consecutive years of hay to one or two of grain.

In regions where summer fallow is practiced, wise land use involves a minimum of tillage and keeping stubble and crop residue on the surface. Large open areas should be avoided, since they allow high wind velocities. This may be accomplished by alternating strips of fallow and strips of stubble or growing crops.

## *Tillage*

Tillage practices which result in a rough cloddy surface, preferably with trash cover or stubble mulch, help check wind erosion. Rough tillage has four main advantages:

1. It catches saltating particles.
2. It decreases the velocity of wind at the surface of the soil.
3. The clods and crop residue prevent soil particles from breaking up into small particles which will drift.
4. It increases the percolation rate and decreases runoff.

The stubble mulch is particularly needed on sandy soils where a rough surface cannot be maintained.

Soils which are susceptible to wind erosion should be plowed as infrequently as possible. Where a one-way disk tiller or a wide sweep tiller is available, it should be used in place of the moldboard plow for summer fallow, fall plowing, and even seedbed preparation. A duck-foot cultivator, tandem disk, or disk harrow will kill weeds and leave a trash mulch if used properly. Many farmers follow the moldboard plowing operation with a subsurface pocket or basin tiller to roughen the surface.

The one-way disk tiller is valuable for partially turning under combine stubble and other crop residue. It leaves a rough surface with a stubble mulch which is ideal for fallow or for seedbed preparation. On fallow land the one-way tiller should not be used more than once a season, because it will pulverize the soil and bury the crop residue too much.

The tandem disk can be used to control weeds and at the same time leave an adequate trash cover for erosion control. It has the added advantage of producing an ideal seedbed which often needs no further preparation.

**Fig. 5-31.** Subsurface tillage implement. The wide sweeps of this implement till the soil and control the weeds while leaving a trash mulch on the surface. (*USDA*)

**Fig. 5-32.** Subsurface tillage. The implement shown in Fig. 5-17 in action. Most of the stubble is left on the surface; therefore the wind cannot erode this soil. (*USDA*)

During recent years several subsurface-tillage implements have become available. Most of these cultivate 3 to 5 inches below the surface, leaving the crop residues on the surface. The most common instrument of this type is the field cultivator equipped with duckfoot shovels or 10- to 12-in. sweeps. The wide duckfoot shovels are effective in killing weeds while leaving the surface relatively undisturbed.

Fig. 5-33. Emergency wind-erosion control. This emergency listing will temporarily protect the soil from wind erosion. (*USDA*)

Large sweeps (20 to 60 in. wide) are available for attachment on the moldboard plow in place of the share and moldboard.

There seems to be no decrease in yield from subsurface tillage as compared to plowing on most soils in dry regions if the fields are kept clean of weeds and drifting is controlled. The principal advantage of subsurface tillage is the control of erosion through crop residues left on the surface. The skill of the farmer can be measured by the amount of cover he can leave without permitting weeds to grow. One to two tons of plant residue per acre, if well anchored in the surface of the soil, is sufficient to control most wind erosion. If fallow land does not have sufficient trash cover, it should not be subjected to the drifting hazards of the fall, winter, and spring periods without additional protection. Such fields should be seeded to a cover crop, listed, or roughened in fall or early spring, before erosive winds occur.

Surface creep may be almost eliminated and saltation greatly reduced as a result of ridging erodible soil. The higher the ridges, the more effective they are in stopping surface creep and saltation and in bringing wind-resistant clods to the surface.

When soil starts drifting, farmers sometimes resort to ridging or listing erodible soil. It is preferable to list the whole of the affected area, and where time does not permit listing at 2-ft intervals, it should be done at 4-ft intervals. If necessary, the spaces between the lister furrows can be listed later. Whatever emergency tillage is done should be wholly effective in stopping erosion. If not, the entire effort will be wasted, for erosion will increase as time goes on.

On some farms the offset-mounted basin disk and the basin tiller are coming into use. They leave the surface rough and are effective in checking drifting on fall-plowed or fallow soil where the soil has become very loose. Stubble or crop strips may be used for the same purpose.

### *Plant Nutrition*

A vegetative cover is the best-known protective measure against wind erosion. Soils which will not support a complete vegetative cover are potential wind-erosion hazards. Failure to produce a vegetative cover in wind-erodible soils is most often caused by a lack of water; however, in some cases there are shortages of essential elements.

Owing to the low rainfall and high evaporation in the Great Plains of North America there has been little if any loss of plant nutrients by leaching. A layer of calcium carbonate is often found in the subsoil. Most wind-erodible soils are light colored and low in organic matter and nitrogen.

With the exception of the sandy soils, field and pot experiments with commercial fertilizer have not indicated a general lack of any of the essential elements in the Great Plains, and the yield is usually directly dependent on moisture supply. In years of favorable moisture conditions, nitrogen fertilization often results in yield increases, especially on land that has been cropped for a long time.

The main value of fertilization in wind-erosion control is to get plants to germinate quickly and to grow beyond the seedling stage, so that they serve to reduce the wind velocity near the ground. Plants that are well supplied with nutrients make more efficient use of the available water.

### *Water Management*

The general cropping practice on much of the prairie area of North America includes a summer fallow at frequent intervals. In the districts where periods of drought are frequent, the land is fallowed every second or third year to conserve moisture. Thus part of the precipitation of the period between crops is stored for use by the succeeding crops. Weed growth is controlled to conserve moisture, for weeds are heavy users of water.

Intertilled crops have been tried as a substitute for summer fallow to overcome the loss of one year's crop. This practice has not been satisfactory, for the intertilled crops used practically all the available moisture, thus defeating the main purpose of the summer fallow. Experiments indicate that from one-fifth to one-third of the summer rainfall is conserved by summer fallow [20]. The amount of water conserved varies from year to year and is determined by the amount and intensity of rainfall, weed growth, soil type, and cultural practices.

Even though summer fallow serves to conserve moisture, it leaves the soil insufficiently covered to resist wind erosion, because the exposed soil particles at the surface are dried out and are easily detached.

Such practices as ridging, leaving strips of standing stubble, and the use of snow fences to trap and hold snow increase soil moisture as the snow melts and also reduces wind erosion.

Contour furrows or contour ridges have been used to reduce runoff and increase infiltration. On an experimental field at Spur, Texas, a system of terraces is arranged so that runoff water from the uppermost terrace is caught by the next terrace and conducted across the field in the opposite direction. In this way water must cross the slope several times before running off the field. This practice, called "water spreading," has very substantially reduced runoff, increased infiltration, and reduced wind erosion by promoting the growth of a more vigorous crop [19].

## SPECIAL WIND-EROSION CONTROL MEASURES

### *Mulching*

The maintenance of crop residues on or near the surface reduces evaporation and constitutes an essential phase of water management in potentially wind-erodible, arable soils, and such maintenance is the greatest single factor in wind-erosion control. The use of mulches for water-erosion control is discussed earlier in this chapter. Many of the same advantages and disadvantages of mulch tillage exist when mulch is used as a wind-erosion control measure.

There are, however, fewer mechanical problems connected with the use of mulch in semiarid regions than there are in humid regions. Most mulch-tillage tools function with little difficulty in dry soil, because the mulch material will slip by without catching on the tillage tool.

The threshold velocity for movement of undecomposed crop residues, even if scattered on the surface of the ground, is higher than for most of the erodible soil particles [11]. If the wind is not too strong, some of its force is absorbed by this material, and erosion of the soil is reduced considerably. If, however, the wind velocity is great enough to remove the crop residue, the threshold value of the affected surface is reduced, and erosion continues at a lower wind velocity than before. To be fully effective in reducing wind erosion, part of the mulch material should be anchored in the soil. The portion which protrudes above the surface forms a very effective trap for particles which move by saltation or surface creep. Another great advantage of mulch tillage in dry-land farming is that the snow will stay where it falls instead of blowing away.

### *Strip Cropping*

In semiarid and arid regions, where most of the serious wind erosion is found, profitable agriculture depends on the use of periodic fallow to store moisture for the use of succeeding crops. It is on these fallow areas that soil drifting first becomes serious. Some farmers in southern Alberta noticed that the last place to drift was on the western part of their fields and that the prevailing winds were generally from a westerly direction. These observations soon led to a division of the fields into alternate strips of fallow and grain, a practice that was so helpful that it has been widely adopted. Today many thousands of acres are regularly cropped in straight strips perpendicular to the prevailing wind.

The minimum width of strips must be the maximum distance that a particle can move in a single leap of saltation plus a reasonable safety

**Fig. 5-34.** Wind strip cropping. Alternate strips of grain and fallow protect this Wyoming field. (*USDA*)

factor. The maximum width will depend on soil and wind characteristics [56]. Strips are seldom wider than 300 ft, and strips of approximately 200 ft are most common. The actual width selected is usually a multiple of the width of the farmer's equipment or some width that would divide evenly into the width of the area to be stripped.

Strip farming has some disadvantages, but if soil drifting is a problem, the inconveniences are outweighed by the benefits. The tendency of soil to ridge at the windward edge of the grain strip is one of the most objectionable features. It may be overcome by proper cultural methods. When it is encountered, the practice of breaking down the stubble with a disk on the windward side of the stubble strip for a distance of 1 or 2 rods is often employed. This permits any drifting soil from the fallow area to distribute itself over this disked portion of the stubble strip, and no abrupt ridge is formed. Under extreme conditions it may be necessary to move the strips of summer fallow and crop gradually to the leeward side across the field.

Strip farming does not eliminate soil drifting, but if the fallow strips are protected by stubble mulching, the wind-erosion hazard is practically eliminated.

### *Windbreaks*

Windbreaks are often employed to aid in the control of wind erosion. They may also be used to provide shelter for farm buildings, control snow drifts, or decrease loss of soil moisture. Windbreaks are common sights on muck land, on sandy areas, and around farmsteads of

**Fig. 5-35.** Windbreaks for fields and farmsteads. A combination of windbreaks for fields and for farmsteads makes this Oklahoma district both more productive and more pleasant for the farm families. (*USDA*)

many areas, such as parts of Russia, China [12], Australia, Africa, and the Great Plains of North America [26]. The main purpose of a windbreak is to serve as a barrier which interrupts and slows down the natural flow of the wind as it passes over the area needing protection. The effectiveness of any obstacle in lowering the surface velocity of wind depends on the shape, height, length, and density of the obstacle and upon the velocity of the wind. The evaluation of the individual effects of these variables has not been solved completely; however,

**Fig. 5-36.** Windbreak on muck land. A row of willows protects a muck field in northern Indiana from erosion. (*J. C. Allen and Son*)

effective windbreaks can be obtained, if they are properly designed and maintained.

Some farmers have used paper screens and cloth wind barriers with only fair success. In other cases lath fences have been used, but the most common windbreaks are composed of trees and shrubs.

Windbreaks should be planted perpendicular to the direction of the prevailing wind, and if the purpose is to protect farm buildings, the trees should be 100 to 150 ft from the center of the building group [1]. To protect a field which is bounded by a highway, the wind break should be 150 ft away from the center of the highway in order to avoid accumulation of snow on the road. For protecting mineral soils in arid regions, from one to a maximum of nine rows of trees is recommended. The plantings should consist of one to three rows of tall trees with dense shrubs on each side. Tree rows should be spaced 10 to 14 ft on irrigated land and 12 to 18 ft apart on dry land. Within the row, trees should be 10 to 12 ft apart, with the exception of the outermost row, which is planted to dense shrubs or trees and should be 2 to 6 ft apart, depending on the variety of plant used [55]. It is especially important to select species which are adapted and which have desirable growth characteristics. A mixture of species is a safety factor.

For the outside row of a tree windbreak in the Northern Great Plains, the following are recommended: Siberian pea tree (*Caragana arborescens*), Russian olive (*Elaegnus angustifolia*), common chokecherry (*Prunus virginiana*), Tatarian honeysuckle (*Lonicera tatarica*), red mulberry (*Morus rubra*), and lilac (*Syringa vulgaris*). The next row should be planted to conifers such as red cedar (*Juniperus virginiana*), Rocky Mountain juniper (*Juniperus scopulorum*), or also to spruces (*Picea* spp.) or pines (*Pinus* spp.). The center row or rows may consist of taller trees like green ash (*Fraxinus pennsylvanica*), box elder (*Acer negundo*), hackberry (*Celtic occidentalis*), honey locust (*Gleditsia triacanthos*), catalpa (*Catalpa* spp.), or cottonwood (*Populus deltoides*).

In a typical east-west windbreak in Nebraska the following species are used beginning with the windward (southern) side:

First row: red cedar (*Juniperus virginiana*)
Second row: western yellow pine (*Pinus ponderosa*)
Third row: western sand cherry (*Prunus bessey*)
Fourth row: Russian olive (*Elaegnus angustifolia*)
Fifth row: cottonwood (*Populus deltoides*)
Sixth and seventh rows: Siberian elm (*Ulmus pumila*)

Specific suggestions on the use of trees and shrubs under various climatic conditions are given in the *U.S. Department of Agriculture Yearbook of Agriculture* on "Trees," 1949.

Windbreaks should be as long as possible, ending in depressions or on sod to minimize the cutting effect of the wind as it sweeps around the ends. Otherwise the same types of dense shrubs that are used for the outside row should also be used at the end.

Properly designed windbreaks will decrease the wind velocity on the leeward side for a distance up to 20 times its height and for about 2 times its height on the windward side [55]. Figure 5-37 illustrates the protection offered by a windbreak.

Such a barrier interrupts and deflects the wind, generally resulting in a decrease in its velocity. However, higher flow rates are obtained

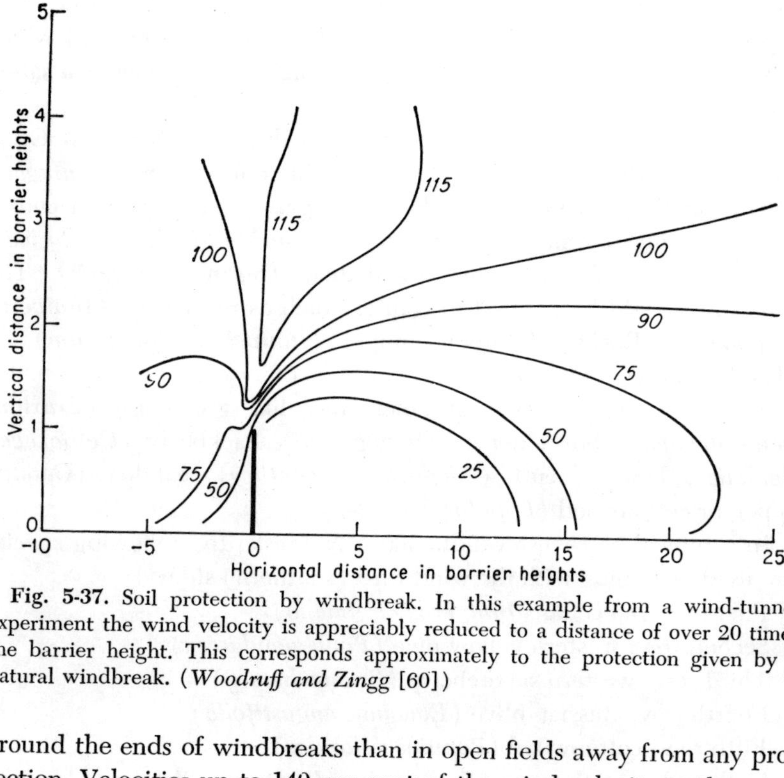


**Fig. 5-37.** Soil protection by windbreak. In this example from a wind-tunnel experiment the wind velocity is appreciably reduced to a distance of over 20 times the barrier height. This corresponds approximately to the protection given by a natural windbreak. (*Woodruff and Zingg* [60])

around the ends of windbreaks than in open fields away from any protection. Velocities up to 140 per cent of the wind velocity in the open have been measured through gaps and around ends of windbreaks [60]. Therefore it is essential to keep the windbreak growing well and to replace dead and dying trees. Above all, it is absolutely necessary to keep livestock out of windbreak areas.

The retardation of wind starts a whole chain of favorable climatic influences. In the vicinity of an effective windbreak, evaporation will be lowered, relative humidity will be higher, and moisture will accumulate because of the retention of snow on the fields. All of these lead to increased crop yields, except in the immediate vicinity of the trees.

It is interesting that in China, where tree shelter belts are more common than in North America, single rows of trees are used almost exclusively. The belts are placed at very narrow intervals (60 ft for willow plantings), and the underbrush is trimmed to allow the wind to pass through. This prevents soil accumulation against the trees if any soil movement should occur, avoids air stagnation and excessively high temperatures near the ground, and gives more room for the growth of other crops.

The planting of trees alone will not provide a solution to soil drifting in dry areas. However, windbreak plantings carried out practically and intelligently in combination with suitable cultural methods help materially in reducing wind erosion. Unfortunately, many areas where wind erosion is serious are too dry for tree plantings other than around the farmstead, where watering is possible.

### *Sand-dune Stabilization*

Active sand dunes are common along sea- and lake shores and in the lower rainfall areas of the world. The total acreage involved is tremendous. Since this area potentially could contribute to the food and timber supply of the world and since the shifting sand endangers crops in adjacent areas, the stabilization of sand dunes is an important part of soil conservation.

Sand dunes consist largely of medium to coarse sand which moves almost entirely by saltation and surface creep. Stabilization depends on reducing wind velocity at the surface or establishing a vegetative cover. There are no practical means of reducing wind velocity, except through the use of windbreaks. But these are difficult to establish in loose, dry, infertile, shifting sand.

Most attempts at control of sand dunes have aimed at complete vegetation with trees or grasses. Any plant which is successful in binding sand must be one that is well anchored, able to hold sand, and catch the wind-borne load. Even more important, it must keep its aerial portion above the accumulating sand and withstand abrasion. Such plants usually have either deeply buried rhizomes or prostrate growth with stems anchored by abundant adventitious, fibrous roots.

Among the sand-binding grasses which have been successfully used,

European beach grass (*Ammophila arenaria*) and prairie sand seed (*Calamovilfa longifolia*) are outstanding. Sand-blowout grass (*Redfieldia flexuosa*) is of importance in the Great Plains of North America. In the Pacific Northwest and in the Great Lakes region of the United States, thick-spiked wheat grass (*agropyron dasystachyum*) is used.

**Fig. 5-38.** Anchored sand. Sea lyme grass (*Elymus mollis*) is an example of vegetation used to stabilize drifting sand. (*USDA*)

Yellow lyme grass (*Elymus flavescens*), sand dropseed (*Sporobolus cryptandrus*), and alkali sacaton (*Sporobolus airoides*) are found growing on inland dunes and are useful in the control of these dunes [31, 51].

The best results in initial stabilization of dune areas can be obtained by use of healthy vegetative material. Sowing seed usually ends in failure.

Woody plants are required to supplement other vegetation in permanent stabilization of sand dunes. However, they are most easily established after the sand is completely stilled. They are usually native species known to be well adapted. Quick-growing species are preferred. Western sand cherry (*Prunus besseyi*) is common on sand dunes in dry areas. Pine trees have proved useful in more humid regions.

It is advisable to begin tree plantation on the windward side and to continue year by year to plant more trees to the leeward as the older growth provides protection. Also mulching with brush during the period of tree establishment is helpful.

### *Stabilization of Muck Land*

Because of the low density of soil organic matter and because of the formation of small lightweight aggregates of saltation size, muck is extremely susceptible to wind erosion. Wind erosion starts on dry, loose muck land when the wind velocity reaches 15 to 18 miles per hour.

Effective wind-erosion control can be accomplished by continuous close-growing vegetation or by the use of windbreaks. On agriculturally used muck land a complete vegetative cover cannot always be maintained. Therefore windbreaks are the most successful method of control.

Among the many types of windbreaks employed on muck land are multiple rows of trees, single rows of trees, periodic grain strips, paper strips, burlap strips, wood-lath snow fences, and other plants and shrubs of various kinds. Paper, burlap, snow fences, and many of the small shrubs which have been used are expensive and difficult to maintain. A wood-lath snow fence can be an effective windbreak, even though it causes considerable air turbulence, if it is placed at frequent intervals across the field. It should be securely fastened to posts and placed in the field immediately after the crop is planted.

When grain crops such as rye, barley, and oats are planted in strips to protect other crops, the strips must be very close together. The wind will bend grain over, and the crop will not assume an erect position while the wind is blowing. Barley strips 12 in. high have been observed to bend over to less than one-half the original height. Consequently, when they are needed, their zone of influence is greatly reduced. When grain is used to protect such crops as onions on muck land, each row should be protected by a grain strip. It is necessary to remove the grain strips at the proper time in order to prevent competition with the main crop. This is expensive and time-consuming.

Multiple-row windbreaks of trees are effective in reducing wind erosion, but they are expensive to establish and require a great deal of land. To be useful windbreaks should be less than 1000 ft apart [17].

One of the most effective and economical methods of controlling wind erosion on muck land is the use of single-row green-willow (*Salix amygdalina*) windbreaks at intervals of 500 to 600 ft [56]. Willow-tree windbreaks cause wind to be deflected up and to filter through. The swaying action of trees offers great resistance to the force of wind. Willow windbreaks are economical to use. The cuttings are low in cost, easily established, and provide quick, effective protection if given cultivation and care. Willow cuttings 18 in. long and ¾ to 1.5 in. in diameter are collected when the trees are dormant and stored until planting time. Cuttings are pushed down into the muck soil with the buds pointing up, leaving 2 to 3 in. of cutting above the surface. They should be cultivated at least two years and trimmed on one side after the trees are 10 ft tall. Only one side each year should be trimmed to maintain the effectiveness of the windbreak.

By proper trimming and by placing windbreaks along field borders the space required may be held to a minimum. Adequate protection may be obtained by using less than 5 per cent of the land. Even though the space occupied by windbreaks on muck land represents a net loss in income, most farmers recognize that the entire cost and the loss of income from land occupied by windbreaks may be paid for by one year's protection to a high-cash crop. On the other hand, such a humid atmosphere may be created on the lee of windbreaks on muck that fungal crop pests become a problem.

## REFERENCES

1. Bates, C. G.: The windbreak as a farm asset, *USDA Farmers' Bull.* 1405, 1944.
2. Beasley, R. P., and L. D. Meyer: A new terrace construction technique, *Agr. Eng.*, **38**:32–36, 1957.
3. Borst, H. L., and R. Woodburn: The effect of mulching and methods of cultivation on runoff and erosion from Muskingum silt loam, *Agr. Eng.*, **23**:19–24, 1942.

4. Brehm, C. D., and H. E. Malmsten: Contour furrows on pasture and range land, *J. Soil and Water Conservation*, **9**:111–114, 1954.

5. Browning, G. M., and F. M. Milan: The lateral movement of water in relation to pasture contour furrows, *Soil Sci. Soc. Am. Proc.*, **5**:386–389, 1940.

6. Browning, G. M., R. A. Norton, E. V. Collins, and H. A. Wilson: Tillage practices in relation to soil and water conservation and crop yields in Iowa, *Soil Sci. Soc. Am. Proc.*, **9**:241–247, 1944.

7. Browning, G. M., and R. A. Norton: Tillage practices with corn and soybeans in Iowa, *Soil Sci. Soc. Am. Proc.*, **12**:491–496, 1947.

8. Browning, G. M., R. A. Norton, and C. K. Shedd: Mulch culture in relation to soil and water conservation and corn yields in Iowa, *Soil Sci. Soc. Am. Proc.*, **8**:424–431, 1943.

9. Carter, L. S., and G. R. McDole: Stubble-mulch farming for soil defense, *USDA Farmers' Bull.* 1917, 1942.

10. Chepil, W. S.: Dynamics of wind erosion: I. Nature of movement of soil by wind, *Soil Sci.*, **60**:305–320, 1945.

11. Chepil, W. S.: Dynamics of wind erosion: II. Initiation of soil movement, *Soil Sci.*, **60**:397–411, 1945.

12. Chepil, W. S.: Wind erosion control with shelterbelts in North China, *Agron. J.*, **41**:127–129, 1949.

13. Chepil, W. S.: Seasonal fluctuations in soil structure and erodibility of soil by wind, *Soil Sci. Soc. Am. Proc.*, **18**:13–16, 1954.

14. Chepil, W. S.: Factors that influence clod structure and erodibility of soil by wind: Organic matter at various stages of decomposition, *Soil Sci.*, **80**:413–420, 1955.

15. Clark, M. W., and W. R. Tascher: Conserving soil by contour farming, *Missouri Agr. Extension Service Circ.* 399, 1939.

16. Cook, R. L., and F. W. Peikert: A comparison of tillage implements, *Agr. Eng.*, **31**:211–214, 1950.

17. Den Uyl, D.: Windbreaks for protecting muck crops and soils, *Purdue Univ. Agr. Expt. Sta. Circ.* 287, 1943.

18. De Roo, H. C.: Subsoiling, plowing, and deep placement of lime or fertilizer in one operation, *Agron. J.* **48**:476–477, 1956.

19. Dickson, R. E., C. E. Fisher, and P. T. Marion: No runoff and no erosion for twenty years, *Texas Agr. Expt. Sta. Progr. Rept.* 1065, 1947.

20. Doughty, J. L., and staff: Report of investigations of the Swift Current, Saskatchewan Soil Research Laboratory, published by the Minister of Agriculture, Ottawa, Canada, 1943.

21. Duley, F. L., and J. C. Russell: Effect of stubble mulching on soil erosion and runoff, *Soil Sci. Soc. Am. Proc.*, **7**:77–81, 1942.

22. Ellison, W. D., and O. T. Ellison: Soil erosion studies: Part VI, Soil detachment by surface flow, *Agr. Eng.*, **28**:402–408, 1947.

23. Free, E. E., and J. M. Westgate: The control of blowing soils, *USDA Farmers' Bull.* 421, 1910.
24. Free, G. R.: Stubble-mulch tillage in New York, *Soil Sci. Soc. Am. Proc.*, **17**:165–170, 1953.
25. Free, G. R.: Investigations of tillage for soil and water conservation: I. A comparison of crop yields for contour vs. up and down slope tillage, *Soil Sci. Soc. Am. Proc.* **20**:427–429, 1956.
26. Gamble, S. J. R., T. W. Edminster, and F. S. Orcutt: Influence of double-cut plow mulch tillage on number and activity of microorganisms, *Soil Sci. Soc. Am. Proc.*, **16**:267–269, 1952.
27. Gardner, H. H., and E. Freyburger: Grass waterways, *USDA Leaflet* 257, 1949.
28. George, E. S.: Spacing distance for windbreak trees on northern Great Plains, *USDA Circ.* 70, 1948.
29. Hamilton, C. L.: Terracing for soil and water conservation, *USDA Farmers' Bull.* 1789, 1943. (Revised.)
30. Hamilton, C. L.: Terrace outlets and farm drainageways, *USDA Farmers' Bull.* 1814, 1939.
31. Hitchcock, A. S.: Controlling and reclaiming sand dunes, *USDA Bureau of Plant Industry Bull.*, 57, 1904.
32. Hockensmith, R. D., and J. G. Steele: Classifying land for conservation farming, *USDA Farmers' Bull.* 1853, 1943.
33. Hockensmith, R. D., and J. G. Steele: Recent trends in the use of land-capability classification, *Soil Sci. Soc. Am. Proc.*, **14**:383–388, 1949.
34. Ireland, H. A., C. F. Sharple, and D. H. Eargle: Principles of gully erosion control in the Piedmont of South Carolina, *USDA Tech. Bull.* 633, 1939.
35. Jepson, H. G.: Prevention and control of gullies, *USDA Farmers' Bull.* 1813, 1939.
36. Joffe, J. S.: The pedologic aspects of working the land: I. Plowing, *Soil Sci. Soc. Am. Proc.*, **10**:446–450, 1945.
37. Karstens, G. A., and R. O. Cole: Terracing for erosion control in Indiana, *Indiana Agr. Extension Service Bull.* 288, 1950.
38. Kidder, E. H., R. S. Stauffer, and G. A. Van Doren: Effects on infiltration of mulches of soybean residues, corn stover and wheat straw, *Agr. Eng.*, **24**:155–159, 1943.
39. Kohnke, H., and A. R. Bertrand: Fertilizing the subsoil for better water utilization, *Soil Sci. Soc. Am. Proc.*, **20**:581–586, 1956.
40. Lunt, H. A.: The use of woodchips and other wood fragments as soil amendments, *Connecticut Agr. Expt. Sta. Bull.* 593, 1955.
41. McCalla, T. M.: Influence of biological products on soil structure and infiltration, *Soil Sci. Soc. Am. Proc.*, **7**:209–214, 1942.
42. McCalla, T. M., and F. L. Duley: Disintegration of crop residues as influenced by subtillage and plowing, *J. Am. Soc. Agron.*, **35**:306–315, 1943.

**43.** Moody, J. E., J. H. Lillard, and T. W. Edminster: Mulch tillage: Some effects on plant and soil properties, *Soil Sci. Soc. Am. Proc.*, **16**:190–194, 1952.
**44.** Nichols, M. L., and J. F. Reed: Soil Dynamics: VI: Physical reactions of soils to the moldboard surfaces, *Agr. Eng.*, **15**:187–190, 1934.
**45.** Peel, T. C.: Influence of mulches on runoff, erosion and crop yields, *S. Carolina Agr. Expt. Sta. 55 Ann. Rept.* 1943, pp. 30–32.
**46.** Ramser, C. E.: Terracing farm lands, *USDA Farmers' Bull.* 997, 1918.
**47.** Ramser, C. E.: Grassed waterways for handling runoff from agricultural areas, *Agr. Eng.*, **24**:412–416, 1943.
**48.** Russell, J. S.: We know how to prevent dust bowls, *J. Soil and Water Conservation*, **10**:171–175, 1955.
**49.** Seamans, A. E.: Recommended practices for soil erosion control, *Montana Agr. Expt. Sta. Circ.* 190, 1948.
**50.** Soehne, W.: Einige Grundlagen fuer eine landtechnische Bodenmechanik, *Grundlagen der Landtechnik*, **7**:11–27, 1956.
**51.** Smith, O. W., H. D. Jacquot, and R. L. Brown: Stabilization of sand dunes in the Pacific Northwest, *Wash. State Coll. Agr. Expt. Sta. Bull.* 492, 1947.
**52.** Spain, J. M., and D. L. McCune: Something new in subsoiling, *Agron. J.*, **48**:192–193, 1956.
**53.** Stallings, J. H.: Effects of contour cultivation on crop yield, runoff and erosion losses, *USDA Mimeographed Circ.*, 1945.
**54.** Stallings, J. H.: Review of terracing data on crop yield, runoff and soil loss, *USDA Mimeographed Circ.*, 1945.
**55.** Staple, W. J., and J. J. Lehane: The influence of field shelterbelts on wind velocity, evaporation, soil moisture and crop yield, *J. Agr. Sci.*, **35**:440–453, 1955.
**56.** Thorfinnson, M. A.: Wind erosion control, *Univ. Minn. Agr. Expt. Sta. Bull.*, 235, 1948.
**57.** Tower, H. E., and H. H. Gardner: Strip cropping for conservation and production, *USDA Farmers' Bull.* 1981, 1954. (Revised.)
**58.** Van Doren, G. A., R. S. Stauffer, and E. H. Kidder: Effort of contour farming on soil loss and runoff, *Soil Sci. Soc. Am. Proc.*, **15**:413–417, 1950.
**59.** Verma, A. B. S., and H. Kohnke: Effects of organic mulches on soil conditions and soybean yields, *Soil Sci.* **72**:149–156, 1951.
**60.** Woodruff, N. P., and A. W. Zingg: Wind-tunnel studies of fundamental problems related to windbreaks, *USDA SCS-TP*-112, 1952.
**61.** Yaeger, H. J., L. S. Robertson, H. Kohnke, and I. D. Mayer: Terracing costs and problems in western Indiana, *Purdue Univ. Agr. Expt. Sta. Bull.* 583, 1952.

# 6

# *Special Soil-conservation Problems*

## WATER CONSERVATION

While soil is an essentially unrenewable resource, water is never lost permanently. Water may be in short supply at a certain place and time, but it will become available with the next rain. The amount and the frequency of precipitation is determined largely by broad geographic and physiographic conditions and only to a very minor extent, if at all, by man's activity. Because water is essential in all agricultural and industrial processes and is basic for life itself, we are greatly concerned about its proper management. Soil conservation and water conservation frequently go hand-in-hand and require the same measures.

### *Soil Moisture and Ground Water*

The water that is held in the soil is called "soil moisture." It is usually under tension, except for short periods after excessive rain or irrigation. Water under tension is able to move only if it is attracted by a greater tension. This means that a dry soil will soak up moisture from a wet soil when they are in contact; a plant root can absorb water from a moist soil as long as the soil is wetter than the "wilting point"; and water can evaporate from the soil to the atmosphere as long as the vapor pressure of the soil water is greater than that of the water in the atmosphere.

Under conditions of general agriculture it is desirable to have the soil at the end of winter at "field capacity" down to the maximum depth the roots will reach. In subhumid and semiarid regions this only happens in depressional areas. There is not enough precipitation during the winter to bring the upland soil to this moisture condition. Even in humid sections not enough water will enter the ground if the soil is too impervious to permit the precipitation water to infiltrate and percolate and allows it to be lost as runoff.

On the other hand, soil should be dry enough during the growing period for enough oxygen to enter to supply the respiration needs of plant roots and microbes. To satisfy these two requirements (to take in most of the precipitation water and to allow excess water to drain away), the soil must be porous. Consequently, a soil that is in the ideal condition for water management is automatically in ideal condition for soil conservation. It will be moist enough to be stable and not to wash or blow away easily, and it will be dry enough to have storage capacity for the next rain. A variety of physical determinations exist that can tell whether the soil is near the ideal condition or not. These include determinations of pore-size distribution and aggregate stability throughout the profile and of the tension with which the moisture is held at various times during the year.

Water that enters the ground but is not held by it percolates to lower depths and contributes to ground water. In contrast to soil moisture, ground water is not held under tension but is free to follow gravitational attraction. Because of the great resistance to flow in the fine crevasses of the subsoil and rock material, the surface of the ground water is not level over larger areas. Ground water is of direct use to crop plants only where it is relatively near the surface. It is the main source of water for domestic and industrial purposes and is frequently used for irrigation. Obviously, therefore, the availability of large amounts of ground water is essential for the economy of a region. Since surface runoff does not contribute to ground-water storage, or does so only to a small extent, it is desirable from the point of view of ground-water management to have most precipitation water enter the soil instead of running off. This shows that the aims of ground-water management as well as those of soil-moisture management are identical with those of soil conservation.

Ground water is generally much more valuable than river water because of its greater purity. In passing through the soil, solids and microbes are filtered out, and only soluble salts are contained in the ground water. The ions found in largest amounts are $Ca^{++}$, $Mg^{++}$, $K^{+}$, $Na^{+}$, $Fe^{++}$, $HCO_3^{-}$, $SO_4^{=}$, and $NO_3^{-}$. The concentrations vary greatly from region to region. River waters contain some of the same salts plus varying, and sometimes very plentiful, amounts of soil. Not all ground water is pure. In limestone areas surface water can penetrate to the ground water through solution cavities without passing through soil and rocks.

## *Flood Control*

A creek or river is in flood stage when its water flows over the banks and covers the bottomland. This bottomland is called "flood plain,"

**Fig. 6-1.** Flood damage. The power of a river in flood stage and the havoc wrought by it is illustrated by this debris the Ohio deposited during one of its great spring floods. (*USDA*)

and its soil material, "alluvium." Floods are caused by excessive amounts of precipitation and by snow melt. They are sometimes aggravated by sediments that have been deposited in the stream beds, thus reducing the channel capacity. The water in streams originates both from ground-water flow and from surface runoff. Where improper agricultural practices have decreased the infiltration capacity and the percolation capacity of the soil, large portions of the precipitation water run off superficially and quickly enter the streams. Water that enters the soil takes quite some time to pass through the ground and to enter a stream directly or to appear on the surface as a permanent or wet-weather spring. Therefore ground-water flow and seepage flow can contribute appreciably only to a flood of long duration, while surface runoff is responsible for flash floods. Rainstorms of excessive

**Fig. 6-2.** Flooded farm land. A 4-in. summer rain caused this headwater flood in Nebraska. Much damage results to the crops, especially if the water remains on the ground for several days. (*USDA*)

rates occur usually in the summer. They do not last very long and do not cover large areas. Therefore flash floods are typical for small watersheds and for the summer season. An example of this is given in Table 6-1, which shows the seasonal occurrence of floods in the Ohio River Basin.

Fairly heavy rainfall lasting several days in succession happens during the winter and early spring. It can spread over huge areas, causing floods in large river systems. Since soil-conservation measures are aimed at getting the water into the ground, they are effective in reducing small-area summer floods that are largely caused by surface runoff. For the same reason, however, soil-conservation treatments increase ground-water flow and have only little effect in reducing floods caused by long-continued rainstorms. The value of soil conservation measures in such a case lies in slowing down the rate at which the water reaches the streams; instead of running off superficially it percolates through the soil and comes to the surface farther

*Table 6-1. Seasonal Occurrence of Floods in the Ohio River Basin, According to Watershed Size*

| Drainage area, sq miles | Maximum annual floods occurring in: | |
|---|---|---|
| | May–September, % of time | October–April, % of time |
| 1 | 99 | 1 |
| 10 | 87 | 13 |
| 100 | 66 | 34 |
| 1,000 | 26 | 74 |
| 10,000 | 10 | 90 |
| 100,000 | 5 | 95 |

From Floods and the farmer, Soil Conservation Service, Milwaukee, Wis., July, 1951.

downhill. If the rain lasts for several days, there is not very much difference in stream flow, whether conservation measures have been used or not. The definite advantages of soil-conservation techniques in reducing any flood are the increased surface storage and the protection from excessive sedimentation. This keeps the stream channels at full capacity and allows greater flow rates before the water spreads into the bottomland.

In some areas floods are beneficial because they rejuvenate the flood plain with fine-textured soil and plant nutrients. Generally, however, they do a great deal of harm. Much coarse soil is deposited over better soil, thus impoverishing the bottomland. The water itself, if it covers the soil for more than a few hours, cuts off the oxygen supply from the soil and consequently suffocates the crop plants and many microbes. This is particularly serious if a flood occurs in the growing season when the microbial activity and the oxygen requirement of the crop roots are high. Other damages caused by floods have been discussed in connection with sedimentation.

More spectacular, but probably less important economically, are the damages to human habitations and industrial plants that are caused by floods. Today, where rivers have lost much of their importance as traffic arteries, there is little reason to establish homes and factories in flood plains. A water supply can usually be found above the flood line.

In addition to soil-conservation measures, two main methods are used for flood control: levees and dams. Levees are placed next to streams in order to confine the flood flow to a restricted portion of the bottomland and to protect the rest of it from inundation and from sediments.

In some cases, levees only protect the flood plain from erosion and deposition but are built with an open end. This permits the floodwater to back around them and to spread over the bottomland. This type of "headwater levee" has the advantage of confining the main flow in the stream channel and yet providing a safety valve for the excess water. Since the floodwater backs up on the flood plain, it does relatively little damage. It has no erosive energy and contains only finer soil particles.

Dams are used in many locations with the purpose of reducing flood flow. This is not the only function which dams have to perform. Some of them serve to store water for cities and industries, to produce electric energy, for irrigation, for recreation, or to act as a settling basin for eroded soil. Dams that are designed for flood control exclusively are built as "dry dams" (Fig. 6-3). The stream water is allowed to flow through them when it is in ordinary stages. During flood flow the gates are partially closed, and the excess water is stored behind the dams. In this type of dam the entire storage capacity can be used for flood control. Consequently, they are most effective for the size used. Since there are no other benefits from these dams, their expense must be charged to flood control alone, while multiple-purpose dams provide income through sale of water, electricity, or through other uses of the water. The value of dams for flood control decreases in proportion to the degree to which they are kept filled. Consequently, farm ponds are of little use in controlling floods.

Since water behind a dam moves very slowly, most of the erosional debris is deposited there, thus diminishing its storage capacity. To maintain the usefulness of a flood-control dam, erosion should be reduced to a minimum in the watershed draining into it. This is particularly important, because there usually are only very few places along a stream that can be used as dam sites and because the area that is covered with water is frequently the best agricultural land.

**Fig. 6-3.** Dry dam. Dover Dam, a concrete structure across the Tuscarawas River in Ohio, was built to help control waters. (*Muskingum Conservancy District*)

## *Farm Ponds*

**Purpose and Value of a Pond.** Farm ponds are built to store runoff water for many purposes. The water is used for livestock, for irrigation, for orchard spraying, for fire protection, for raising fish, and for boating and swimming. It is also of advantage for wildlife and for domestic waterfowl. They are of only limited value for flood protection, since they are kept nearly full, and for ground-water gain, because the pond bottoms must be essentially impervious to keep the water from draining away. In order to keep excessive sedimentation from reducing its capacity, the watershed draining into the pond has to be protected from erosion. The pond itself provides little erosion control, except through the use of a protected spillway and sodded waterway leading down from the pond.

**Requirements for a Good Farm Pond.** A farm pond should be located in a natural draw or on a hillside where enough good water can be expected to flow into it. For reasons of economy it is best to

**Fig. 6-4.** Farm pond. This large farm pond (6½ acres of water) in Illinois is protected from livestock by a fence and ½ mile of multiflora rose hedge. Four thousand five hundred pine trees have been planted to decrease wind velocity and to add beauty. (*USDA*)

place the dam at a point where the least amount of earth movement is necessary. The size of a farm pond can vary from ¼ acre up. Ponds that are to serve as water supply for irrigation have to be fairly large. The pond when filled should be at least 6 feet deep in 20 per cent of the area, and there should be as little shallow water as possible, to avoid excessive evaporation and to prevent the entire pond from freezing up. This would be disastrous to fish life, to the water supply for the stock, and to fire protection. Shallow water also encourages weed growth.

The size of the watershed above the pond should be large enough to keep the pond filled with water most of the time, yet not so large that there is frequent runoff through the emergency spillway. The ratio between the watershed area and the pond area depends on soil, climate, topography, land use, soil-conservation measures, and the average depth of the pond. It should be large enough to keep the pond filled through extended dry periods and small enough to avoid excessive siltation and frequent flow over the emergency spillway. In the Middle West this ratio is between 6 to 1 and 20 to 1. A wider ratio is used only if erosion in the drainage area is under control.

In the interest of having clear water flowing into the pond it is desirable to have much of the watershed in erosion-resistant land use, such as forest, pasture, or meadow. It is particularly important to have the lower reaches of the watershed just above the pond in sod, so that erosional debris is filtered out instead of being carried into the pond. Heavily eroding areas and barn lots should never be in the watershed above a farm pond.

Since evaporation can be the cause of serious water losses, coniferous trees should surround the pond at a distance in order to reduce the wind movement over the pond. Deciduous trees are not desirable, because their leaves falling into the water would impair its quality. In no case should any trees be grown on the dam or near the spillway, because their roots would create seepage danger.

A pond should only be located on a fairly tight soil, since losses from seepage are excessive where the pond bottom is pervious and where shale or rock are near the surface. It is impossible to get a watertight seal between rock and soil.

**Individual Parts of a Farm Pond.** A farm pond consists of a reservoir that is to store the water, a dam, a mechanical spillway, an emergency spillway, a stock watering trough below the pond, and a fence to protect the pond from livestock [17]. The dam should have an impervious clay core that penetrates into the solid earth underneath and on the sides of the dam. The dam itself should be at least 6 ft wide at the top, if it is 10 ft high [7]. It should be 1 ft wider for each additional 5 ft in height. The slope toward the water is usually about 3 to 1, while the slope away from the water can be 2 to 1 [9].

A mechanical spillway is placed into the pond so that its intake is about 2 ft lower than the emergency spillway. It consists frequently of a concrete box that is connected to a pipe which leads under the dam and discharges the water in a spillway outlet made of concrete. From here the water is carried off in a sod waterway. Sometimes the same pipe is used to provide water to a stock tank, but usually a special water pipe with a filter at the entrance serves for this purpose. A float valve regulates the flow of water into the tank.

In order to avoid water overtopping the dam, an emergency spillway of large cross section must be provided for every farm pond. It is usually a sodded waterway placed on the side of the dam leading into

**Fig. 6-5.** A well-maintained farm-pond dam. Reed canary grass protects the dam from wave action and the drop-inlet spillway carries off excess water safely. (*USDA*)

a well-protected area below. Occasionally, the emergency spillway is built of masonry or concrete, especially in larger ponds. Its overflow elevation is between 1 and 3 ft above the intake of the mechanical (pipe) spillway and about 2 ft below the top of the dam. The fence around the pond should be considered an integral part of a farm pond, since it is needed to protect the pond from livestock.

**Maintenance of Farm Ponds.** There are many causes why a farm pond may not maintain the desired water level. Some of these are an inadequate spillway that erodes to form a deep gully through the dam, seeping past the water pipe, a pervious core or pervious bottom, tree roots or animal burrows in the dam, erosion of the dam, plugging up of the pipe spillway—sometimes as a result of frost—or trampling by animals. Whatever the cause of potential failure, it is always advisable to be constantly vigilant in order to anticipate required maintenance work. Probably the most important item is to protect the

**Fig. 6-6.** Fertilizing a farm pond. Where plant nutrients in the soil are low, fertilization of farm ponds boosts fish production by increasing the growth of microscopic plant life. (*USDA*)

emergency spillway from erosion, but any of the other causes can be responsible for unsatisfactory performance of the pond.

A special feature of farm-pond operation is the raising of fish [8, 12, 13]. Properly managed, a farm pond can yield more meat per acre than a pasture. Such management includes the planting of the right species of fish, the maintenance of enough water throughout the year, the protection of the water from mud and excessive growth of weeds and algae, the establishment of a desirable level of plant-nutrient elements in the water, and last but not least, the fishing out of excess numbers of fish, so that there will be enough food for the remaining fish to grow large.

The protection of the pond, the dam, and their immediate surroundings from erosion of the watershed draining into the pond and from livestock is an important item of pond maintenance.

## AREAS OF EXCESSIVE EROSION HAZARD

Soil movement on land that is used for crop production represents the bulk of man-induced soil erosion. The problems connected with soil conservation on these areas are so general that control measures have been developed that are applicable under most conditions. Unless other methods suffice to protect the soil, a change to a less erosive land use can solve the problem. This has been discussed in previous chapters.

In some specific cases erosion hazards are particularly large because of intensive use or excessive steepness or because of constant exposure to large volumes of running water. The land use generally cannot be altered, and most of the conventional erosion-control measures for agricultural land do not apply. Examples of such cases are barnyards, stream banks, roadsides, airports, and the spoil banks of worked-out mines.

### *Barnyards*

Barns and barnyards are frequently placed on knolls or hillsides in order to have them on dry ground and sometimes to provide a second-story entry into the barn. The combination of fairly steep slopes with excessive traffic of livestock and machinery makes barnyards sometimes the most severely eroded part of the farm. On some farms roof water is allowed to flow freely into the barnyard and contributes to the erosion hazards. In barnyards wind erosion generally causes serious soil losses only on sandy ground. In semiarid areas deposition of sand in the barnyard is a greater problem than erosion.

To minimize barnyard erosion several measures can be used. The farmyard should be located on a well-drained soil, and any water that might tend to flow into it from higher locations should be diverted. The buildings should be located so that most of the traffic will be on the contour. Where steep lanes cannot be avoided, they should be hard-surfaced. Feeding lots should be concreted. This will not only control erosion but it will be better for the animals' health and will pay for itself through saving feed and manure. All buildings should be provided with downspouts, and the water should be discharged into underground tiles or some other type of safe outlet. This may incidentally serve to supply a cistern with soft water.

**Fig. 6-7.** Soil dunes in a farmstead. Soil eroded by wind in the surrounding fields has been deposited in this South Dakota farmyard with disastrous results. (*USDA*)

Dispersing livestock from the barnyard will help to control erosion. Use of movable sheds and placing fenced-in haystacks in fields can help to avoid unnecessary concentration of cattle on the farmyard. For reasons of sanitation it is advisable to place hogs and poultry on new ground at least once each year.

Wind-erosion problems on a farmyard can be solved by planting a windbreak at some distance around it. The trees of the windbreak must be protected with a fence from livestock. Windbreaks provide the additional advantages of a more pleasant climate in the farmyard,

**Fig. 6-8.** Rehabilitated farmstead. Picture of the same farmstead as shown in Fig. 6-7 taken eight years later. Wind-erosion control in the fields and plantations of trees and shrubs in the farmyard have restored the usefulness of the farmstead and have added beauty to it. (*USDA*)

**Fig. 6-9.** Windbreak. A double row of pine trees eliminates the wind-erosion hazard in this northern Indiana farm garden. (*J. C. Allen and Son*)

of reduced fuel needs for heating the house, and of added beauty to the immediate area.

## *Stream Banks*

Stream banks present a special erosion problem because the changing water level leaves much of the bank bare of vegetation and therefore very vulnerable to erosion [1]. The large amount of water flowing

**Fig. 6-10.** Stream-bank erosion. A whole row of walnut trees gives way to the swollen waters of the Santa Clara River in California. (*USDA*)

in a stream supplies a great deal of energy, especially during storm periods. Alluvial soils in headwater areas usually contain little clay and therefore are not very cohesive and erode easily. The most severe erosion occurs on the outside of bends in the stream, where the water hits the bank with its full force. One of the periods of severe erosion is the winter, when ice floes hit the banks and when constant freezing and thawing loosen the soil from its compound.

In the case of stream erosion it is particularly difficult to distinguish between man-induced erosion and geologic erosion. Stream erosion can be severe even though man has not interfered with the hydrologic conditions of the watershed. Where agriculture has taken the place of the original land cover, runoff peaks are higher and more frequent and aggravate stream-bank erosion. But regardless of the extent to which this damage is man-induced, it requires effective control measures, since stream-bank erosion can cause extensive losses of valuable bottomlands. It is also responsible for a large proportion of the sediment in streams and reservoirs.

Stream-bank erosion control requires some of the following measures: the water should flow the fastest in the middle of the stream, and along the banks conditions should be favorable for vegetation. Where the erosion hazard is particularly great, mechanical devices may be needed. The best method of stream-bank erosion control is the decreasing of runoff peaks, thus decreasing the eroding energy and keeping the water level more uniform. In small watersheds this may be possible by the use of soil-conservation practices; on large ones only storage dams can accomplish a material reduction of flood peaks.

In order to have the water flow fastest in the middle of the stream, any obstruction should be removed. Sometimes tree trunks or other large debris float down the stream when it is in flood stage and are deposited when the water subsides. Branches, leaves, and stones pile up behind such obstructions and force the water toward the bank.

Whenever practicable it is advisable to straighten out meandering streams, since erosion is particularly severe at the outside curves where the current strikes the banks. All trees on the inside curve should be cut near the ground or grubbed out by the roots, as they tend to direct the stream to the outside curve, causing more erosion.

Vegetation is essential in protecting stream banks, but not all vegetation is desirable. The vegetation near the water's edge should be adapted to occasional submerging. It should be able to bend over to let floating debris pass. Trees hold such debris; thus a barrier is created and the effective cross section of the stream is reduced. The water piles up behind this barrier and flows with increased intensity, causing damage where it strikes the bank. However, judgment must be used to avoid creating excessive velocities.

The majority of grasses cannot withstand longer periods of flooding. Moreover, their root systems are not deep and strong enough to resist the eroding effect of fast-flowing stream water.

Reed canary grass (*Phalaris arundinacea*) is one of the few grass species that thrives near water and can survive considerable flooding. A plant that has proved to be particularly effective in protecting

Fig. 6-11. Protected stream bank. Riprap made from treated fence posts that are tied to stakes in the bank of the irrigation canal in Colorado keeps erosion under control. Tree branches and brush were piled between the fence and the bank for further protection. (*USDA*)

stream banks is the purple-osier willow (*Salix purpurea*), a low-growing shrub. It bends over to let large debris float past, yet it is firmly anchored into the ground and can withstand the eroding effect of a stream once it is fully established. Another shrub useful for this purpose is the red-osier dogwood (*Cornus stolonifera*) [15].

Stream banks are frequently nearly vertical because of undercutting by the water. Before any other erosion-control measure can be started, the bank has to be sloped to a 1 to 1, or preferably a 1 to 2, gradient, so that vegetation can grow on it. Where the erosion hazard is severe, the slope has to be protected with rocks, stakes, and brush (Fig. 6-11). Rocks are generally used at the water's edge. Stakes are driven into the ground in order to protect newly established vegetation. Sometimes living stakes are used. They soon sprout and develop roots and become firmly established. Only shrub plants should be used for such stakes, since any tree would have to be cut out later. Covering the newly planted or seeded stream-bank slopes with a mulch of branches is helpful. The use of alder branches is recommended, especially in early fall when they contain ripe seed. Occasionally it will be necessary to cover the ground with poles or fence wire to protect it from erosion.

Maintenance of stream banks requires deflecting the flow from the banks, removing any obstruction in the stream that would cause the water to cut the banks, keeping livestock from tramping on them, and repairing quickly any damage that may occur. Where a stream flows through a pasture, it should be fenced out from the animals with wire or a hedge of multiflora rose (*Rosa multiflora*) [5]. If the stream water is needed for the livestock, the access places should be protected with rocks. Any trees encroaching on the stream should be cut.

### *Roadsides*

In the construction of roads an effort is made to have their grades as smooth as possible in order to have more economical, safe, and pleasant traveling. This requires cutting through the hills and filling the valleys. Raw subsoil and even "soil-parent material" and rocks are exposed. In the effort to keep roads well drained, ditches are placed on the sides of the road that catch the water from the road itself and from the land above it. In hilly areas such ditches can easily become gullies unless erosion is controlled. The berm represents another danger spot. Cars frequently are driven on the berm in emergencies or for a rest, and they compact the soil and injure the vegetation. The wind created by traffic makes establishment of seedings on the berm difficult, especially when the ground must be firm enough to bear vehicles.

In addition to sheet and gully erosion on the roadsides, landslides are frequently encountered in areas where deep cuts have removed the support of the earth above the road. Deposits from roadside erosion may endanger traffic and make road maintenance costly, or they may cover adjacent farm land.

The control of roadside erosion can best be accomplished with close-growing, low vegetation, generally grasses and legumes [10]. The establishment of such sod-forming plants requires that the slopes of all banks be no steeper than 2 to 1 (50 per cent). For reasons of maintenance with power machinery, however, it is better to have the grades be only 3 to 1 (33 per cent). Since such slopes, especially the south-facing ones, are exposed to great temperature variations and severe drought, it is advisable to cover the freshly graded roadsides with mulching material, such as straw. This is also very necessary to protect them from the impact of the raindrops and from severe erosion. Frequently the use of mulch on road banks means the difference between establishment of a protective sod and the washing off of seed and soil. Where the slopes are steep or on the berm or where the wind of the traffic might remove the straw, spraying of a small amount of asphalt is effective in stabilizing the mulch. The sooner a sod cover can be obtained, the better. But as road construction goes on throughout the year, fresh roadsides will frequently be exposed for several months before seeding can be done and the grass itself can give erosion protection. It is therefore advisable to apply a mulch immediately after construction and not to wait for the seeding.

Whenever water from higher ground has the tendency to flow over the road banks, it should be diverted and carried off to a safe outlet. If it has to cross the road, a concrete drop inlet may be needed to conduct it into a culvert. At some sites in road ditches, chutes and drop spillways offer good solutions.

Grasses are the best type of vegetation for road-bank protection, but some low-growing legumes such as crown vetch (*Coronilla varia*) or wild flowers may be useful under some conditions. The grasses should be sod forming and should spread by underground stocks. Bunch grasses are not desirable. Only perennial grasses are of value. Quick cover can be obtained from small grains, Sudan grass (*Sorghum sudanense*), common rye grass (*Lolium multiflorum*), and other annuals, but care has to be taken that these do not compete too severely with the more desirable species.

Some of the grasses that are useful in roadside erosion control in the United States are Kentucky bluegrass (*Poa pratensis*), Canada bluegrass (*Poa compressa*), creeping red fescue (*Festuca rubra*), Chewings fescue (*Festuca rubra* var. *commutata*), tall fescue (*Festuca arundinacea*), smooth brome (*Bromus inermis*), quack grass (*Agropyron repens*), colonial bent grass (*Agrostis tenuis*), redtop (*Agrostis*

**Fig. 6-12.** A drop-inlet box. This box in combination with a large culvert protects the road from flood damage and erosion. (*Portland Cement Association*)

*alba*), perennial rye grass (*Lolium perenne*), Bermuda grass (*Cynodon dactylon*), carpet grass (*Axonopus affinis*), centipede grass (*Eremochloa ophiuroides*), buffalo grass (*Buchloe dactyloides*), blue grama grass (*Bouteloua gracilis*), crested wheat grass (*Agropyron cristatum*), western wheat grass (*Agropyron smithii*), Indian rice grass (*Oryzopsis hymenoides*), sand dropseed (*Sporobolus cryptandrus*), and weeping love grass (*Eragrostis curvula*).

These are adapted to the various climatic and site conditions of the country. In addition to these species, reed canary grass (*Phalaris arundinacea*) is of value in controlling erosion in roadside ditches.

Kudzu (*Pueraria thunbergiana*) is used successfully to cover road banks in the Southeastern United States. It grows so aggressively, however, that the danger exists of crowding out trees and other vegetation in adjacent land unless it is regularly cultivated.

**Fig. 6-13.** Road-bank stabilization. In the Southern United States kudzu provides perfect protection of roadside ditches and banks as well as of gullies. (*USDA*)

Roadside cuts expose subsoil, which is usually low in available plant nutrients. To ensure success in establishing vegetation on road banks, ample fertilization is a necessity. In fact, on old highway banks fertilization alone may be sufficient to give the ground a dense plant cover. A "complete" fertilizer is best for these purposes. For maintenance, usually nitrogen alone is required. Grass along roads should be clipped two or three times during the year to keep it dense and to control trees and weeds.

### *Airports*

Erosion on airports, airfields, or flight strips can be severe for various reasons. In order to have a nearly level area the soil of the high spots is sometimes removed and placed in the depressions; consequently, subsoil is exposed that has a low infiltration capacity and is difficult to cover with vegetation. Such subsoil is poorly aggregated and easily washes or blows away. Rain water that falls on a concrete runway runs off and creates an added erosion hazard. Wind is frequently severe on airfields, because they must have unobstructed approaches and there is nothing to mitigate the velocity of the wind. In addition, the propeller wash from the airplanes represents winds of hurricane velocity. On airfields that are not properly protected from erosion, dust can be a serious problem [21]. Dust clouds reduce the visibility and have caused many accidents in landings. Dust is also detrimental to engines, sometimes reducing their life span to one-half the normal period.

Erosion control on an airfield requires the employment of the same principles that help to conserve soil elsewhere—a solid ground cover and avoiding concentrations of surface runoff. It is not always easy to bring about such conditions on an airport. Surface water from higher-lying ground should be diverted, and the topography of the airport itself has to be carefully planned so that no accumulation of water occurs. Similar to the situations on barnyards and roadsides a dense grass sod is the most desirable ground cover. Once it is established, it helps to increase infiltration capacity, it slows down surface runoff, and it protects the ground from rainfall impact and from wind. Grass species that spread by underground stocks and form sods like those suggested for the protection of roadsides are also recommended for airports [6, 20].

The establishment of grasses on airports requires as much preparation as that of a lawn. The ground should be tilled carefully and be. fertilized adequately. Seeding rates should be high, and mulch should be used to protect the surface soil from heat and drought. Irrigation helps to assure the establishment of grasses and should be used where feasible. In order to avoid blowing away of straw mulch, it can be partially pushed into the ground with a dull disk or a small sheep's-foot roller. Straw can also be held on the ground by spraying it with cutback asphalt.

Maintenance of a grass sod includes mowing, fertilizing, and—under severe conditions—watering. Sometimes areas of heavy use need to be reestablished, preferably with the temporary exclusion of traffic.

### *Strip-mined Lands*

When coal, iron, phosphate, and other minerals are near the surface, they are mined by "stripping" the overburden from the minerals, thus making them accessible to removal with power equipment. This leaves soil, subsoil, rocks, and any other material that happens to cover the minerals in great ragged mounds. The deeper the minerals are in the ground, the higher will be the mounds and the smaller will be the proportion of the original surface soil in the banks. As these are steep and loose and possess little natural coherence, erosion rates are extreme as long as they are not covered by vegetation. Most of the eroded soil is usually deposited in the excavated openings or in the troughs between the ridges of overburden, and only a portion of it is carried beyond the mined area itself. Nevertheless, erosion can be a serious problem in strip-mined lands, because it slows down formation of a productive surface soil and causes sedimentation in lower-lying areas.

The only effective method of erosion control and the logical way of bringing strip-mined lands back into usefulness is to revegetate them. The specific procedure employed depends on the climate, the physical and chemical nature of the overburden, the topography, and the economic conditions [14, 18, 19].

In arid and semiarid land not much can be done to cover spoils with vegetation. Since land is of little value in such areas, there is no financial inducement to expend any effort to do so. In humid areas revegetation of spoil banks starts by natural means soon after the spoils have been deposited. The first species to grow are those that propagate by wind-blown seed, such as cottonwood (*Populus deltoides*), sycamores (*Platanus occidentalis*), and wild lettuce (*Lactuca scariola*). To what extent and at what rate such revegetation will be complete depends largely on the physical and chemical nature of the material. If it consists exclusively of large igneous rocks, practically no plant will get established. Presence of free mineral acid or toxic concentrations of copper, zinc, and other metal ions will also prevent plant growth. Where spoil banks consist of calcareous glacial till, volunteer vegetation usually becomes established very rapidly. Within three or four years trees, grasses, legumes, and many other species cover such areas almost completely.

Frequently such spoil-bank areas are planted to trees in order to provide an income and also to remove the unsightly banks from view. If physical and chemical conditions are conducive, trees do well on steep spoils. In order to make a higher use of strip-mined land it has to be graded to a maximum slope of 25 per cent so that agricultural implements can be used on it. Excellent growth of meadow and pasture crops has been achieved on graded calcareous spoil. Grading may increase the erosion hazard, at least temporarily, by increasing the size of the watersheds and the distance surface water can flow over unprotected, raw ground. Special care is required to conduct surface runoff to a safe outlet.

## WILDLIFE AND SOIL CONSERVATION

### *Benefits from Wildlife*

Wildlife is a by-product of land use, and in many ways soil-conservation measures increase the numbers and improve the health of wildlife. The monetary income that the farmer gets from wildlife is usually small. Actually, however, wildlife is big business. Many states report incomes of over 100 million dollars per year from transactions directly connected with game and fish [2]. But most of this money is spent on hunting and fishing equipment, clothing, food, dogs, transportation, hunting licenses, and the like, while the farmer reaps little or no direct financial reward. But there are other benefits from wildlife on the farm. Otherwise unusable land can be utilized to supply food for the farmer's table and enjoyment in observing the creatures of nature. Many farmers like to hunt or do favors for their hunting friends [4]. Birds will eat a great many insects that might otherwise damage the crops. About half of all bird species in the United States feed exclusively on insects, yet it is difficult to prove the economic advantage of insect destruction by birds. Insects multiply fast, and it is mostly climatic changes and diseases that keep insect numbers in check. It is equally unlikely that the seed eaters among the birds destroy enough weed seeds to contribute materially to the control of weeds.

In general, the farmer has no great economic inducement to exert any effort to increase the wildlife crop, unless he himself can harvest it profitably or he can rent out the privileges of this harvest, i.e., the hunting and fishing rights. Fortunately he does not have to go to any great trouble to have wildlife on his farm. Natural succession will in most regions bring brush cover to the odd areas of the land that cannot be used for any other crop. The farmer merely has to tolerate it where it is not in the way.

## *Types of Wildlife*

Wildlife is a rather indefinite term unless defined. In this discussion it includes all vertebrates except man and the domesticated animals.

**Fig. 6-14.** Ruffed grouse. Ruffed grouse chick in the Allegheny National Forest in Pennsylvania. The partridge berries provide food for wildlife. (*USDA*)

Of particular interest are the game and fur animals, the game birds, wild waterfowl, songbirds, and fish. But the definition includes also rodents, amphibia, and reptiles. What type of wildlife will occur in a given area will depend on the climate, soil, vegetation, and cultural practices. Soil conservation refers largely to farm land, and hence such areas as mountains, deserts, and large swamps are not considered here. It is quite evident that the species of wildlife on a well-cared-for farm may be considerably different from those on the same land before it was put into cultivation [11]. The difference between farm wildlife, forest wildlife, and wilderness wildlife must be recognized. This discussion is concerned with farm wildlife. In many areas of the world,

**Fig. 6-15.** Deer. Doe and fawn in young jack-pine timber growth in Superior National Forest, Minnesota. (*USDA*)

however, farms are adjacent to forest or wilderness, so that deer, bobcats, or beavers may be found on farms.

## *Methods of Increasing Wildlife Populations*

Whenever it seems desirable to increase the numbers of one or many species of wildlife, various methods are employed. They include hunting restrictions through regulations, predator control, wildlife refuges and parks, artificial replenishment by restocking from game farms, and environmental control. No doubt all these are valuable, but the control of the environment is outstanding in its importance. Unless the surroundings are favorable to a wildlife species, no amount of restocking and protection from predators and the hunter's gun will be very effective.

### *Environmental Requirements of Wildlife*

The type of environment most favorable to wildlife depends largely upon the species considered. A few generalizations, however, can be made. All wildlife requires food and shelter. These should be in close proximity, because the cruising range of some of these animals is small. The type of food varies from plants to insects to vertebrates.

**Fig. 6-16.** Multiflora rose hedge. Multiflora rose provides an effective fence against livestock as well as food and shelter for wildlife. (*USDA*)

Most upland game and upland birds do not need open water, such as ponds or streams, but get enough water in their food and in the dew collected on the plants. The seed eaters among the birds, for instance, the sparrows and the finches, require free water because of their dry food. In densely wooded areas feeding grounds for both birds and browsing animals may be provided by cutting large trees or by using herbicides to reduce shade at ground level, thus permitting a variety of food plants to grow.

Shelter may consist, for example, of a standing field of alfalfa, a hedge, or a forest. Brushy cover that furnishes protection in winter and summer is the principal need. The preference of the animals for the different types of shelter depends on their size and their living habits. Under any circumstances the most desirable shelter is continuous, forming travel lanes and giving easy access to the food areas.

Open water is an obvious necessity for fish, waterfowl, beavers, and several other mammals. It is an attraction also to animals that do not absolutely need it.

### *Soil-conservation Practices that Encourage Wildlife*

It has been said that all a farmer has to do to encourage wildlife is to practice soil conservation. There is much truth in this. Increase in soil fertility produces more food and consequently more and larger animals. Strip cropping provides many field borders, thus giving the desirable edge effect between two forms of vegetation that supply food and shelter. Avoiding overgrazing of pastures and burning and grazing of woodland saves many nesting places for birds and gives shelter to other animals. Windbreaks, fenced-in farm ponds, planted gullies, stream banks, roadsides, and weedy fence rows also help to make a more desirable habitat [3, 16]. Perhaps one of the most important items is that soil-conservation measures result in smaller distances between food and shelter. Also, the more regular flow of springs and small creeks that frequently result from the adoption of soil-conservation farming is a decided asset. Colloidal matter, such as clay and humus, in runoff water is particularly bad for fish. Where improved use of the land results in clearing up the water of the streams, fish increase in numbers.

The concept of land-use-capability classes is another feature of soil conservation. Since it is recognized that wildlife contributes least per acre of all farm crops, the poorest land (land-use-capability class VIII) is designed for wildlife and recreation. This conscious recognition of wildlife as a crop is pointed out to every farmer who has land of this nature, and consequently many nonfarmable areas are planted to thickets and plants that provide food for wildlife during the lean part of the year. It should be remembered that, while wildlife is the main crop on class VIII land, it grows as a by-product on all land.

## *Damage Done by Wildlife*

However well venison may taste, however enjoyable the song of the birds, and however valuable the pelt from the fur-bearing animals, wildlife can do much damage. This is especially true if one species gets very numerous. Rabbits bite off young seedlings of orchard trees and pine plantations. Deer can ruin spruce forests with their antlers. Crows pick whole rows of corn. While foxes eat mice, they also eat chickens. This list can be expanded considerably, but in general it is not difficult to keep all wildlife except rodents down to numbers that cause no serious damage. It has been estimated that rats eat and destroy over 3 million tons of food a year in the United States in spite of extensive control measures.

It is the task of the wildlife management specialists to develop methods that will favor desirable wildlife without letting them get too numerous and to devise control measures for undesirable species.

## REFERENCES

1. Albrecht, W. A.: Drainage ditches can be erosive too, *Soil Conservation*, **16**:141–142, 1951.
2. Allen, D. L.: Wildlife management in land-use problems, *U.S. Fish Wildlife Service, Wildlife Leaflet* 309, 1948.
3. Allen, D. L.: Recent trends in farm wildlife management, *Trans. 4th North Am. Wildlife Conf.*, Washington, D.C., 1949, pp. 253–260.
4. Allen, D. L.: The farmer and wildlife, Wildlife Management Institute, unnumbered bulletin, Washington, D.C., 1949.
5. Anderson, W. L., and F. C. Edminster: Multiflora rose for living fences and wildlife cover, *USDA Leaflet* 256, 1949.
6. Anonymous: Airport turfing, U.S. Department of Commerce Civil Aeronautics Administration, June, 1949.
7. Atkinson, W. S.: How to build a farm pond, *USDA Leaflet* 259, 1949.

8. Ball, R. C., and H. D. Tait: Production of bass and bluegills in Michigan ponds, *Mich. Agr. Expt. Sta. Tech. Bull.* 231, 1952.
9. Beasley, R. P.: Characteristics of farm ponds, *Missouri Agr. Expt. Sta. Bull.* 566, 1952.
10. Brant, F. H., and M. H. Ferguson: Safety and beauty for highways, *Grass: USDA Yearbook Agr.*, 1948, pp. 315–318.
11. Crawford, B. T.: Some specific relationships between soils and wildlife, *J. Wildlife Management*, **14**:115–132, 1950.
12. Davidson, V. E.: Farm fishponds for food and good land use, *USDA Farmers' Bull.* 1983, 1947.
13. Davison, V. E.: Managing farm fishponds for bass and bluegills, *USDA Farmers' Bull.* 2094, 1955.
14. DenUyl, D.: Hardwood tree planting experiments on strip coal mine spoil banks of Indiana, *Purdue Univ. Agr. Expt. Sta. B ll.* 619, 1955.
15. Edminster, F. C.: Streambank plantings for erosion control in the Northeast, *USDA Leaflet* 258, 1949.
16. Edminster, F. C., and R. M. May: Shrub plantings for soil conservation and wildlife cover in the Northeast, *USDA Circ.* 887, 1951.
17. Hamilton, C. L., and H. G. Jepson: Stock water developments: Wells, springs and ponds, *USDA Farmers' Bull.* 1859, 1940.
18. Kohnke, Helmut: The reclamation of coal mine spoils, *Advances in Agron.*, **2**:317–349, 1950.
19. Limstrom, G. A., and G. H. Deitschman: Reclaiming Illinois strip coal lands by forest plantings, *Univ. Illinois Agr. Expt. Sta. Bull.* 547, 1951.
20. Morrish, R. H., A. E. Rabbitt, and E. B. Calc: Airfields and flight strips, *Grass: USDA Yearbook Agr.* 1948, pp. 319–323.
21. Yoder, E. J., and W. H. Daniel: Turf investigations for airports and highways, *Purdue Univ. Eng. Bull.* 78, 1952.

# 7

# *Economics of Soil Conservation*

## THE PLACE OF ECONOMICS IN THE SOIL-CONSERVATION SCHEME

The basic reason for conserving soil is the survival of mankind. The irreplaceable nature of soil does not permit us to have a purely economic viewpoint of soil conservation. Any economic gain that might result from soil conservation is of secondary importance in the general scheme of things, but it gives an important incentive to the individual farmer to adopt such measures. No matter how much he realizes his responsibility as steward of the land, he will only be able to protect the land from erosion and to raise its productivity if his farming operation pays for itself.

Every piece of land has a certain market value that is related to its present and potential productivity. But actually, the intrinsic value of land is much greater than money; it lies in its ability to feed and clothe man for countless centuries. This cannot be expressed in monetary terms.

To determine whether soil-conservation measures are economically sound or not, limits of time and area must be established. Practically all soil-conservation measures benefit the individual farm as well as the area downstream. As a matter of fact, they usually benefit the entire community through increased production and purchasing power. But the farmer can only invest in soil conservation to the extent that he himself is repaid. If the community, state, and nation expect to reap advantages from soil-conservation measures, they should participate in defraying the expense of conservation. This encourages the farmer to use methods that might be of dubious economic value if he had to carry the load alone.

Most forms of agriculture represent a rather slow turnover of money. This is particularly true with soil-conservation measures. The first year a terrace is built, the yields probably will be lower than before because of the large areas of subsoil exposed. Reforestation takes many years to pay for itself. Other improvements, such as proper fertilization and tillage, usually bring a return on the investment within a few months. Generally speaking, a farmer must be able to anticipate economic gains within five or ten years if he is to adopt a soil-conservation measure.

## RELATIVE COST OF SOIL CONSERVATION

### *Costs of Soil Erosion*

Erosion costs are great for the farmer, but they are huge to the nation. The most obvious cost of soil erosion is the lowered productivity of fields and pastures or their complete destruction. But there are many other losses that count in the many millions of dollars annually for the United States. The costs of soil erosion are caused by three main phenomena: erosion itself, increased runoff, and deposition of erosional debris.

**Erosion.** Erosion removes soil and plant nutrients from fields and pastures, and it creates gullies that are expensive to cross and worse to bypass. The productive capacity of an eroded area is lowered, generally in proportion to the amount of erosion that has taken place. Erosion washes out highway and railroad fills, even barns and homes. Erosion pollutes stream water with silt and clay.

**Increased Runoff.** Increased rates and amounts of runoff are frequently the result of erosion, because the absorbent surface soil is removed and the tight subsoil exposed. This lowers the infiltration capacity. Less water can enter the soil, and crops suffer from drought. The greater runoff results in accelerated erosion and in floods. Floods cover valuable farm land with water and frequently drown crops. Practically any building, furniture, machine, or other equipment that is in the path of a flood loses in value or is completely lost. Floods cause also losses of livestock and, in extreme cases, of human lives.

**Deposition of Erosional Debris.** Soil sediments cover fertile valley land with relatively unproductive material; they fill up reservoirs, stream channels, and ditches. They choke culverts and cover fences, roads, and even buildings.

**Farming Eroded Land.** Farming eroded and eroding land is like operating an inefficient industrial plant. It is expensive to till, seed, and harvest an acre that is not fully productive. And it is a waste of money to put fertilizer on the land, only to allow it to wash away.

It must be recognized that this refers principally to land where erosion is a serious factor. Stable soils on flat land or on gentle slopes represent practically no erosion problem. There is little difference between conservation farming and traditional farming on such land.

**Indirect Costs of Erosion.** The indirect costs of soil erosion are too numerous to mention. They include lowered crop yields, smaller livestock production, and the lowered purchasing power of the farmer. Others are the lowering of the water table and the subsequent greater costs for drilling and pumping and the decrease of the wildlife population.

The most serious cost of erosion to a nation lies in the fact that its food-producing potential is reduced.

## *Costs of Soil-conservation Measures*

**Adopting Soil-conservation Measures.** To determine the costs of soil conservation, we need to establish which agricultural improvements are strictly soil-conservation measures and which are merely sound agriculture. It is impossible to draw a definite line. Such specific cases as relocation of fences to conform with the contour, establishing terraces and waterways, and leveling and planting gullies are unquestionably soil-conservation measures. But the use of lime and fertilizer, the tilling of the land, the construction of buildings, and the purchase of seed and livestock can hardly be called anything but ordinary farm operations. Nevertheless, the change-over to conservation farming may involve a substantial increase in the use of soil amendments; it may require the purchase of new tillage implements and many other expenses that would not have been incurred otherwise. It is questionable whether tile drainage, forestation, pasture improvement, building of ponds, and similar operations should be classed as specific soil-conservation techniques or as ordinary farming.

Obviously, the level of soil conservation that is to be reached will largely determine the cost of it. Another problem that faces the farmer who wants to convert to "conservation farming" is the speed with which this is done. A leisurely rate will involve less cash outlay, but on the other hand, it postpones the time when the full benefit of the conversion is realized.

**Cost of Individual Improvements.** A variety of factors determine the cost of the individual soil-conservation measures. Any monetary values refer to only a specific set of circumstances. It seems better, therefore, to classify these measures in groups according to initial expenses and operating costs.

*Low-cost Group.* The group of conservation measures that require low initial expense and operating cost comprise contour strip cropping, wind strip cropping, contour cultivation, mulch tillage, and land-use adjustments. In most cases changing from the original farming methods to these practices involves relocation of fences, but otherwise little expense. After these improvements are established, farming is generally less expensive than before.

*Medium-cost Group.* Terracing, gully repair, and the establishment of sod waterways, while expensive for each area on which these improvements are done, must be considered as medium-cost soil-conservation measures, since the effects extend over a rather large area. Therefore the cost per acre is not very high. In the case of terracing, costs per acre are greatly affected by the steepness and regularity of the slope of the land, the area to be terraced, the type and length of outlets needed, and the type of equipment used. After each of these three improvements has been established, farming is simplified, and maintenance costs are small.

*High-cost Group.* Establishing a new fencing system, necessitated by a change in field boundaries and by the desire to protect woodlands from grazing animals, requires a large expenditure of cash and labor. Actually, the total length of fence per acre of farm is slightly greater on the conservation farm than on the farm using old established methods. The curves that are necessary in contour fences present a tricky problem, and stretched wires may become so short during cold weather that they pull over the posts. On the other hand, straight fences in hilly country go over hills and through ravines and are difficult to build and to maintain. Planting living fences and windbreaks is costly per acre planted, but the maintenance is inexpensive.

Constructing ponds, building a field drainage system with tiles and ditches, and establishing irrigation are important and expensive farm improvements. A well-built drainage system and also a farm pond require little upkeep. Irrigation, on the other hand, is expensive year by year. Probably the most expensive conservation method is the use of fertilizer and lime. This is particularly true because these amendments are used almost every year. Again, fertilization can hardly be considered a specific soil-conservation technique.

Land-use adjustments and increased crop yields brought about by adopting a soil-conservation system of farming frequently create the need for increasing number of livestock and for erecting more buildings.

**Cost of Operating a Soil-conservation System.** Once the change-over to soil-conservation farming has been accomplished, the change in the cost of operating the farm will largely depend on the level of soil conservation and crop production attained and the types of conservation measures introduced. Most of the specific soil-conservation techniques, such as strip cropping, mulch tillage, contour cultivation, terraces, and sod waterways, cause only a slight change in expenses. Costs may go up or down. On the other hand, such general improvements as fertilizing, liming, and irrigating represent a great annual cost. Probably the most important change in operational costs results from land-use adjustments. Generally, the acreage in row crops is decreased, and the acreage in sod crops is increased. This results in an over-all decrease in cost of tillage, cultivation, and harvest and in a more uniform distribution of labor during the year. The increase in yields of field crops requires more harvesting expense per acre, but less per-unit weight of product. It also costs more to maintain the larger numbers of livestock.

### *Returns from a Soil-conservation System*

**Returns from Individual Improvements.** To what extent the adoption of conservation measures will be profitable depends on the nature of the land and on the managerial skill of the operator.

In calculating the returns of a soil-conservation system of farming it is desirable to determine the returns from each individual measure, because the farmer will want to adopt only those that promise to be profitable. However, the operation of a farm is difficult to separate into its financial components. Some aspects lend themselves rather well to financial analysis, for instance, fertilization and irrigation. Others are hard to calculate. How could we determine the economic value of a sodded waterway? The most important difference in returns will result from land-use adjustments. The actual returns for a given situation depend on the price relationships of the main products of the two systems of farming.

**Net Profits of Soil-conservation Farming.** Whether conservation farming or "soil-depleting" farming is more profitable depends on a variety of circumstances. For instance:

The original level of farm management under the soil-depleting system

The practices included in conservation farming

The type of soil, especially its erodibility

The erosiveness of climate and topography

The amount of erosion that has already occurred on the farm

The type of farming system that fits best the physical and economic situation

The time period considered (for a very short time—one or two years—the returns from conservation farming can hardly be expected to pay for the investment)

From all this it is evident that a definite answer to the question of whether changing over to conservation farming promises increased profits or not can only be stated for a given set of conditions. Studies on this subject by Blosser [1], Janssen and Robertson [3], Sauer, McGurk, and Norton [5], and Sauer et al. [4] show that "conservation pays," if a period of several years is considered. Conservation farming requires more labor and capital and more skillful management, and it generally brings more profits.

**Long-time Benefits from Conservation Farming.** The aim in using soil-conservation methods is to maintain and increase the value of the land. By increasing the content of organic matter and plant nutrients of the soil and by allowing more water to enter the soil instead of running off, the soil becomes increasingly productive. Farmers who have used soil-conservation techniques for a number of years report that the yield level has been raised and with it the sales value of the farm, if not in the same proportion.

### *Financing the Change-over*

Generally, a farmer who has worked his land in the old tradition will not have sufficient ready cash to finance an immediate change-over to the conservation system. He can, however, change gradually, "plowing back" his profits from the first conservation methods to pay for the next ones. This may take many years. The success of this type of slow change-over depends largely on the right sequence of adopting the various improvements. The best sequence will not always be the same, but frequently liming, fertilizing, contouring, and the construction of sodded waterways should be the first methods used. The fast change-over to permanent agriculture requires much capital and more labor than is normally available on the farm. The farmer may be forced to borrow a sizable amount of money and hire outside labor. The advantage of the fast change-over is that it will bring returns faster and control erosion sooner.

Whether the farmer decides to adopt soil conservation quickly or at a leisurely pace will depend on his economic status and his personality. The majority of farmers prefer an intermediate approach that allows them to utilize the available farm labor for much of the work but have some of the jobs that require large machinery done by custom operators. Generally, the change-over takes from one to six years.

## OTHER ECONOMIC FACTORS

### *Farm Adjustments Resulting from Soil Conservation*

The type and the extent of adjustments that have to be made as a farm is changed from the traditional system to the conservation system depend on the level of the original farming, on the level of conservation farming to be reached, and on the type of land. Much more of a change is needed on steep and erodible soil than on relatively flat land. In the majority of cases the total production increases materially.

While the acreage of row crops is usually decreased in an effort to reduce erosion, the yields per acre are increased, and the row crops are grown on the better land only. Consequently, the row-crop production per farm may not be greatly changed. Small-grain production may be increased or decreased, depending on the rotation changes that are desirable for the farm. Hay and pasture yields are almost always substantially increased, because both acreage and yield per acre are increased. More hay and pasture mean more livestock, especially of the type that can utilize to advantage large amounts of roughage. Soil-conservation measures result in more cattle, both beef and dairy, and more sheep, but generally not in an increase in the number of hogs.

To store the larger harvests and to provide shelter for the greater numbers of livestock, additional buildings are frequently needed.

### *Tenancy and Soil Conservation*

Tenancy is frequently blamed for increased erosion. It is true that sometimes a tenant is given little inducement to keep up the productive capacity of the land, especially if he is only renting a field. But both landlord and tenant have an interest in conserving the soil. If we analyze a renting contract critically, it appears that the tenant pays rent to the landlord in consideration for the service of the landlord in preserving the farm unit, including the soil, in a usable and productive condition. In other words, it is the right of the tenant to use the land in a manner to make a maximum profit. It is the landlord's responsibility to take measures to interest the tenant in soil conservation. The tenant must, of course, benefit from his soil-conservation work.

If the tenant is the son or some other relative of the landlord, he will expect to reap the results of his efforts by inheriting the farm in a better condition. Tenants who have assurance of staying on the farm for a long period of time see the advantage of investing in soil improvements, but most tenant contracts in the United States are for one year only. Rent contracts in Europe are generally for a longer duration. A tenant is not even interested in liming the land if he assumes that he will not reap the benefit of the liming.

The landlord is responsible for such maintenance or improvement operations, and there should be a provision in every land lease for the tenant to be reimbursed for unexhausted improvements when he leaves the land. There is a saying that the best way for a tenant to move to a better farm is to improve the one he is on. This is true only if he has the cooperation of the landlord and if he is going to stay on the farm for many years. Of course, a landlord will generally appreciate the efforts of a good tenant and keep him as long as possible.

In some cases landlords use special incentive payments for conservation measures by the tenant. An equitable tenant contract may be more effective in bringing about soil conservation than any other single item. Frequently, however, the landlord is not so satisfied with the somewhat lower initial returns of soil-conservation farming as he is with those of land exploitation. It is the responsibility of both the tenant and the landlord to acquaint themselves with the principles of permanent agriculture, so that each can do his part in maintaining the productive capacity of the land [6].

### *Soil Conservation and Taxes*

Since soil conservation is desirable from the viewpoint of the commonwealth, national or local governing bodies should make it financially attractive for the farmer to introduce and use soil-conservation methods. In some cases such inducements have gone so far as to provide tax money as incentive payments for the farmer who adopts specified soil-conservation measures.

This may be necessary, because the financial returns from some conservation practices often are realized only after a number of years.

A sample of tax adjustment to foster soil conservation is the legislation concerning "classified forests." These are wooded areas that are protected by the owner from pasturing and fire. Taxes are assessed on a nominal per-acre value.

Whenever soil conservation increases the gross and net income, taxes are automatically increased. One of the problems of soil conservation and taxation is that land values are based on production, and the good farmer who makes something out of a mediocre piece of land is penalized. An equitable assessment of taxes can be based on a sound and detailed soil survey that would take the productive capacity of the land into account.

### *Soil Conservation and the Market*

**Effects of Market Conditions on Soil Erosion and Soil Conservation.** Extensive farming, as it is found far from markets and under conditions of poor transportation, usually means that only small areas are devoted to high acre-value crops. Much land is left in either forest or prairie, and as soon as a piece of cultivated land begins to deteriorate, it is allowed to revert to natural vegetation and a new piece is selected. Under such conditions of agriculture, erosion is not much of a hazard. In the face of ever-increasing populations and expanding railroad and highway systems this type of farming is fast becoming a thing of the past, but it still exists in central Africa, northern Asia, and other areas that we consider to be on the fringe of civilization.

High prices for a certain product encourage its production. The acreage devoted to such a crop is enlarged, and it occurs in frequent succession in the same field. Such one-crop agriculture is employed particularly where the climate is favorable to this crop or where distance to markets or difficult transportation tends to force the farmer to concentrate on a nonspoiling high-value crop, such as wheat, corn, cotton, or tobacco. Frequently such a crop system results in too much plowing, in loss of organic matter, and therefore in accelerated erosion. Usually a one-crop system is not economical with regard to labor. Most of the work connected with it is done in a few short periods during the year, and not much attention can be given to improvement and maintenance of the soil.

Where market conditions, and especially transportation, become more favorable, it becomes advantageous for the farmer to sell perishable livestock products, fruits, and vegetables. This means a diversified land use. The farmer's time is more fully used, his income per acre increases, and he values his land more highly and is willing to invest more money and labor in the improvement of the soil.

Both the very extensive and the very intensive types of farming cause relatively little erosion. It is the farming of middle economic intensity that is most hazardous for the soil.

**Effects of Soil Conservation on Utilization and Marketing of Farm Produce.** Not only does the market situation influence erosion and conservation; this influence is reciprocal. Introducing soil-conservation measures on a farm usually results in the production of more forage than before and consequently will cause the farmer to increase livestock numbers, especially of the type that is efficient in the utilization of hay and pasture. He may also attempt to feed a larger proportion of roughage to the animals than he did before. This can be accomplished if the roughage is of superior quality and contains high amounts of protein. Since the total yield of row crops per farm is not greatly affected by the introduction of conservation practices but a greater amount is needed to balance the roughage in the ration, a smaller quantity of grain but a larger quantity of livestock products may reach the market. In the face of ever-increasing demands for high-quality food this is a very desirable result of soil conservation.

In a wider sense, soil conservation has a profound influence on the nation's market. In the United States farms furnish the capital for the purchasing power of more than 25 million people. In addition to this, industry depends on the soil for a large proportion of all its raw materials.

Obviously the productiveness of the soil must have a tremendous effect on the economy of the country.

Both the costs and returns of soil conservation are large, and wherever soil conservation has been practiced for a few years, the returns become larger and the costs become smaller. No farmer who has harvested the yields of several season's soil-conservation work ever wants to go back to the old system. He knows that for one or two years he can make more money, but after that the yields decline, the soil washes away, and gullies make farm operations difficult. He continues to use soil-conservation practices because he knows from his own experience that soil conservation is sound economics.

## REFERENCES


1. Blosser, R. H.: How disposition of crops affects the economics of soil conservation, *J. Soil and Water Conservation,* **9**:169–174, 1954.
2. Davis, R. O. E.: Economic waste from soil erosion, *USDA Yearbook Agr.,* 1913, pp. 207–220.
3. Janssen, M. R., and L. S. Robertson: Economic effects of cropping and livestock systems on a rolling central Indiana farm, *Purdue Univ. Agr. Expt. Sta. Bull.* 625, 1955.
4. Sauer, E. L., H. O. Anderson, R. H. Blosser, P. E. McNall, and O. J. Scoville: Conservation problems and achievements on selected Midwest farms, *Ohio Agr. Expt. Sta. Spec. Circ.* 86, 1951.
5. Sauer, E. L., J. L. McGurk, and L. J. Norton: Costs and benefits from soil conservation in northeastern Illinois, *Univ. Illinois Agr. Expt. Sta. Bull.* 540, 1950.
6. Wallace, J. J.: Landlord–tenant cooperation gets conservation on the land, *J. Soil and Water Conservation,* **10**:22–23, 1955.

# 8

# *Farm Planning for Soil Conservation*

## DETERMINING THE RESOURCES OF THE FARM

Many individual methods exist that are designed to reduce soil losses from erosion and to increase the productivity of the land. In order to translate them into an effective program of soil conservation these methods must be viewed from the operation of the individual farm. While several soil-conservation methods transcend farm boundaries, the organization of soil conservation must be based on the economic unit of land operation—the farm. The first step to accomplish this goal is an inventory of the natural resources and the improvements on the farm.

A variety of information is required before a conservation plan can be designed. This includes land and water features, existing cropping practices, livestock and improvements, the climate, the economic environment, and the personality of the farmer and his family.

### *Land and Water Features*

Normally a soil resource map of the farm is prepared to show the following features:

Soil type—basic productivity of the land, texture, fertility, stonyness.

Degree of erosion or of sedimentation.

Slope—steepness, length, choppiness of slopes, location of large gullies.

Water—surface water such as streams, temporary streams, ditches, ponds. Availability of water for livestock use and for irrigation. Some information on the depth to ground water is implied through the name of the soil series.

A farm in west central Indiana is chosen as an example to illustrate the steps in conservation farm planning.

Generally an aerial photograph is used as the basis for such a farm map. A frequently used scale in the United States is, 4 inches on paper corresponds to 1 mile. This means that 1 sq in. represents a 40-acre field. After a map is finished, it is frequently enlarged to double size (8 in. corresponds to 1 mile) for better legibility. The use of aerial photographs has the great advantage that no measurements are required and that the location of soil boundaries and other features can be done much more quickly and accurately than on blank paper, since many items such as trees, roads, fences, and soil differences can be readily seen on aerial photographs. Generally, black and white photographs are used because they are cheaper, but color photographs help greatly in distinguishing ground features and are therefore preferable.

Various techniques can be used in presenting the natural features on the map. A national system of conservation-map symbols has been established in the United States. For local conditions, however, variations from this system are used.

An area in which soil type, slope, and erosion are reasonably uniform is delineated, and a symbol is placed into it. This symbol looks like a fraction with a figure designating the soil type in the numerator and a letter or a figure for the slope and a figure for the degree of erosion in the denominator.

$$\frac{\text{Soil type}}{\text{Slope–Erosion}}$$

The figure designating the soil type is made up of four digits. The first one describes the texture of the surface soil as follows:

1. Clay
2. Clay loam
3. Silty clay
4. Silt loam
5. Loam
6. Sandy loam
7. Sand

The second and third digits denote the catena in which the soil occurs. The fourth digit indicates the internal drainage of the profile:

1. Very poorly drained
2. Poorly drained
3. Moderately well drained
4. Well drained
5. Excessively drained
6. Upland soil profile without development of *B* horizon
7. Slight depression
8. Depression soil
9. Deep depression

The second figure in each soil area gives the per cent of average slope. The third figure represents the erosion condition.

The letters *A* to *G* are often used to designate steepness of slope, or sometimes the average slope is given in per cent (Fig. 8-1).

In one frequently used scheme these letters correspond to the slope ranges shown in Table 8-1.

*Table 8-1*

| Slope symbol | % of slope |
|---|---|
| *A* | 0–2 |
| *B* | 2–6 |
| *C* | 6–12 |
| *D* | 12–18 |
| *E* | 18–25 |
| *F* | 25–35 |
| *G* | Over 35 |

Symbols represent water-erosion conditions in this manner:

*Table 8-2*

| Erosion symbol | Erosion condition |
|---|---|
| + | Recent deposition |
| 1 | Little to no erosion, 0–25% of surface soil gone |
| 2 | Moderately eroded, 25–75% of surface soil gone |
| 3 | Severely eroded, more than 75% of surface soil gone |
| 4 | Very severely eroded, most of surface soil and some subsoil gone, gullied |

Additional symbols are needed for such conditions as detrimental deposits, landslides, and man-made features. More details on mapping techniques are given in the *U.S. Department of Agriculture Soil Survey*

*Manual, Agriculture Handbook* 18. Some of the commonly used symbols are illustrated in Fig. 8-5.

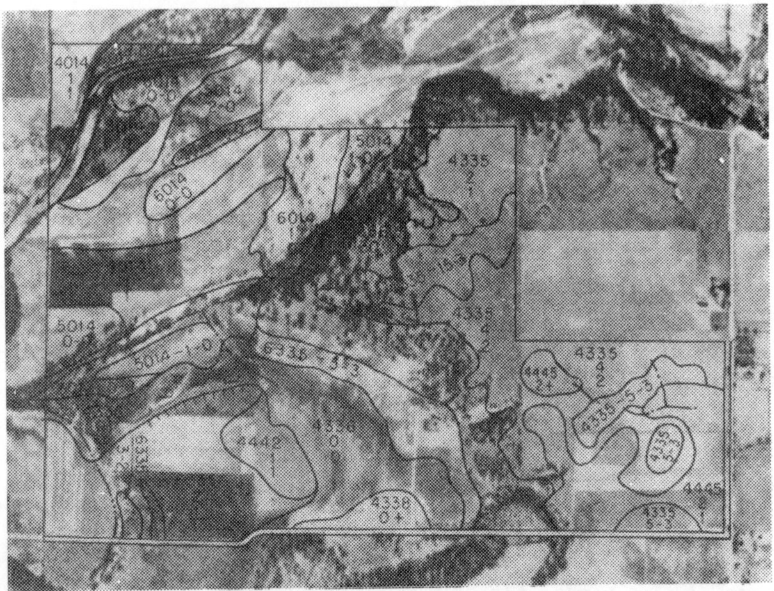


**Fig. 8-1.** Soil resource map. 3012: Eel silty clay loam; 3014: Genessee silty clay loam; 4014: Genessee silt loam; 4335: Fox silt loam; 4338: Westland silt loam; 4442: Sleeth silt loam; 4445: Ockley silt loam; 4666: Rodman silt loam; 6014: Genessee sandy loam; 6335: Fox sandy loam.

The combined information on soil type, slope, and erosion makes it possible to determine the land-use capability class of a piece of ground (Fig. 8-2). In the case of uneroded soils of best quality, the slope groups correspond to the land-use-capability classes as follows:

*Table 8-3*

| Slope group | Land-use-capability class |
|---|---|
| *A* | 1 |
| *B* | 2 |
| *C* | 3 |
| *D* | 4 |
| *E* | 6 |
| *F* | 7 |
| *G* | 7 (woods only) |

Erosion or poorer quality of the soil decreases the land-use capability of the individual slope groups.

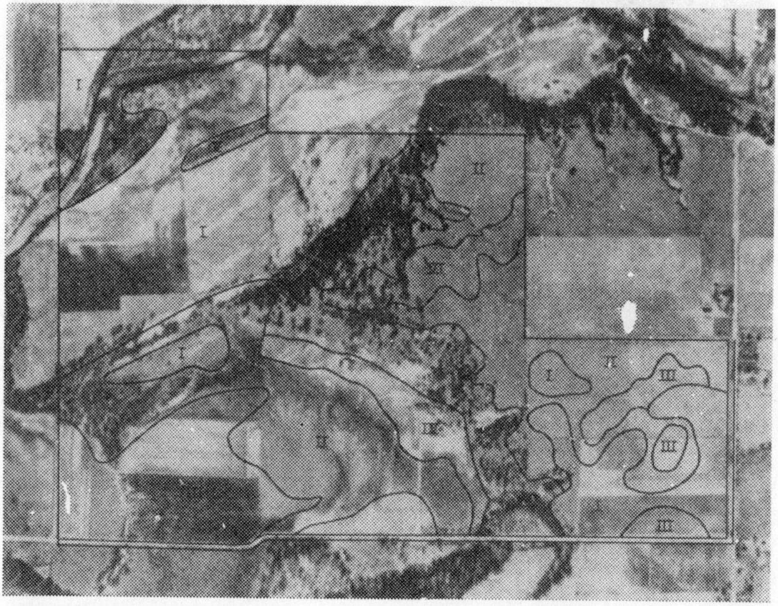

**Fig. 8-2.** Land-use-capability map. Green: class I; yellow: class II; red: class III; blue: class IV; uncolored: class V; orange: class VI; brown: class VII.

### *Land Use, Livestock, and Improvements*

In the preparation of the farm map the present distribution of land use is noted (Fig. 8-3). By the use of a planimeter the acreages of the various crops can readily be estimated. Fences are shown and if possible, the location of drain tiles. Buildings and livestock have to be inventoried in detail.

### *Climate*

In order to be able to know what crops can be grown successfully on the farm, climatic information is needed. This includes annual precipitation, distribution of precipitation over the seasons, sunshine frequency, heat units for the growing season (temperature above a certain minimum for the growth of various crops), relative air humidity (saturation deficit), and length of frost-free period.

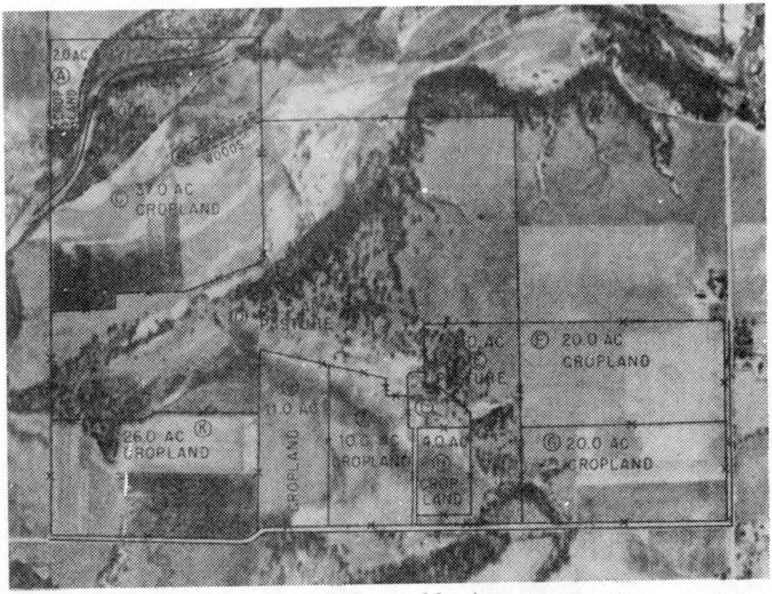


Fig. 8-3. Original land use.

### *Economic Environment*

The distance to market, the nature of the roads, prices of various agricultural products and of farm requirements, purchase and sales organizations, and the financial status of the operator represent part of the economic environment of a farm.

### *The Human Factor*

As important as any other part of the farm in determining what soil-conservation measures will fit the specific case is the personality of the farmer and his family. Age, health, education, and the outlook on life of the members of the family determine to a large extent the nature of the farm operations.

## DETERMINING METHODS OF AGRICULTURE

### *Assigning Land to Its Proper Use*

Once the physical features of the farm have been determined and mapped, it is necessary to classify all land according to its proper use. By proper use is meant obtaining the highest economic return from the land compatible with soil conservation and the maintenance of productivity. The method of land-use-capability classification used by the U.S. Soil Conservation Service has been discussed in detail in Chap. 5.

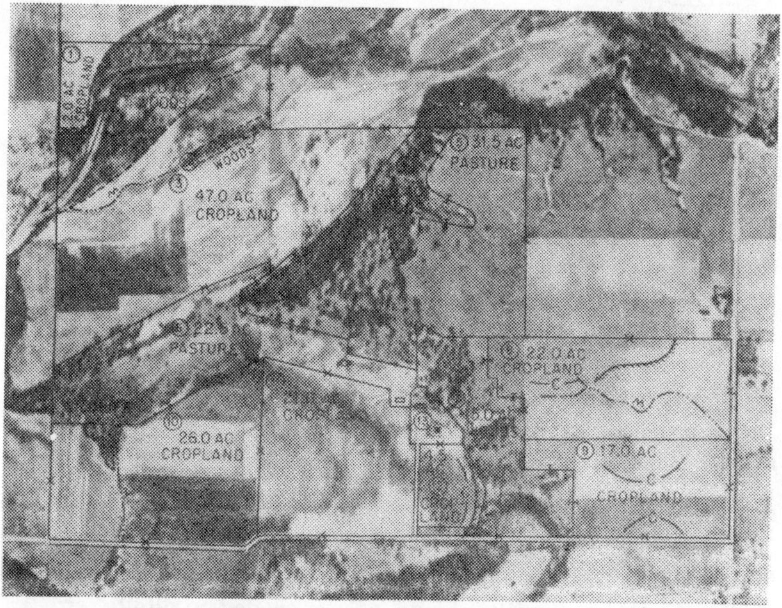


**Fig. 8-4.** Proposed land-use map.

The greatest value of classifying the land according to its use capability is to indicate at a glance to the farmer and land planner the maximum intensity of agricultural use that can be practiced safely.

An examination of the land-use-capability map of an Indiana farm (Fig. 8-2) will show that some very good land was not being used to its maximum capacity. Two rather large areas of class I land in field *D* (Fig. 8-3) were in unimproved pasture. Class II land in field *E* and class III land in field *D* were also only used for pasture. It may be seen (Fig. 8-4) that these areas were easily converted to more intensive use by relocating a few rods of fence, when a conservation plan was prepared for this farm.

A land-use-capability map also directs attention to those areas that are being used too intensively. Fields *D* and *E* on Fig. 8-3 have class VII land that is heavily wooded, yet it was grazed by livestock. On the new farm map which takes into account the capability of the soil, most of this area is excluded from pasturing and left to produce timber for the farm's lumber needs and for sale.

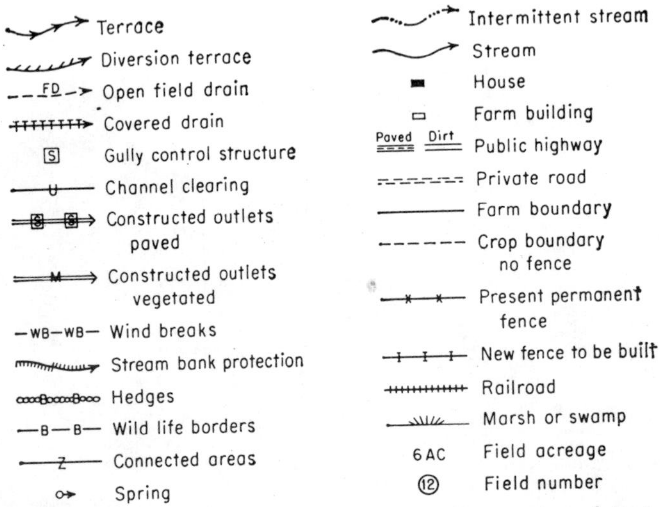


**Fig. 8-5.** Map symbols commonly used by the U.S. Department of Agriculture.

In the selection of the crops for a given piece of land soil erosion is not the only consideration. A droughty soil, for instance, fits a winter grain better than a spring crop. A poorly drained soil, even if no erosion hazard exists, should not be seeded to alfalfa, etc.

## *Choice of a Cropping System*

The establishment of the proper land-use group for every area on the farm does not in itself determine what specific crops to use. In the case of the first three land-use classes the choice of cropping systems is large and frequently difficult. The greater the erosion hazard, the smaller will be the proportion of row crops and the greater the proportion of sod crops. Methods have been worked out to estimate what amounts of erosion may occur under various conditions of soil type, steepness and length of slope, and rotations and other specific soil-conservation practices. Such estimates permit the choice of a combination of these factors that will restrict erosion to "permissible" rates. How to arrive at the best crop sequence for a given case is discussed in Chap. 4.

On land suited for limited cultivation the choice of cropping system is restricted to small grain and sod crops. Usually meadow is maintained as long as it gives satisfactory yields. The land is then plowed or only cultivated, small grain is seeded, and a new meadow established in the grain.

In case of the four land-use classes "not suited for cultivation" the only crops are pasture plants or forest trees. The climatic and soil conditions will usually determine which species of plants should be used.

### *Fertilization*

A vigorously growing crop that produces high yields and covers the ground quickly and thoroughly requires a soil well supplied with nutrients. It also removes more plant nutrients than an average crop. For this reason special attention has to be given to ample and proper fertilization and liming. Some soil-conservation measures call for extra plant food. Mulch tillage requires larger amounts of nitrogen than plow culture. Deepening the root bed is successful only if the subsoil is fertilized. The increased amount of soil water resulting from the various conservation treatments causes leaching of substantial amounts of plant nutrients and on the other hand raises the plant food requirements because of more abundant growth.

Fertilizing is a special branch of soil science that is too involved to be discussed here in any detail. The principle to be followed in soil-conservation farming is to raise the soil to a high plant-nutrient status and to maintain it there by replacing what is lost by cropping or any other cause.

### *Arrangement of Fields*

Once a decision is made as to what rotation or rotations should be used on a farm, it will be necessary to distribute the crops in such a manner that approximately the same amount of yield is harvested from each crop. This means that about the same acreage should be devoted to a given crop each year, assuming that all land used for this crop is of approximately the same productivity. Therefore in case of a four-year rotation of corn-soybeans-wheat-meadow, for example, the land for this rotation will have to be divided into four approximately equal fields. Of course, each of these fields does not necessarily have to be in one continuous piece. The actual location of the individual fields on a farm is not of great importance, except that minimum lengths of fence and lanes should be required and all fields should be readily accessible. In hilly land it will also be advisable to make field boundaries coincide as much as possible with contours.

In planning the farm shown in Figs. 8-1 and 8-2, an attempt was made to adhere to the principles stated above; however, it was necessary to compromise in certain cases. In keeping with the operator's wishes, the fields were made as large as possible to permit the efficient use of farm machinery. All interior fences were removed from field 3 (Fig. 8-4) to simplify the production of continuous row crops in this rich alluvial soil, where few erosion hazards exist. Fields 9 and 12 are farmed as one unit, thereby producing four nearly equal-sized fields from fields number 8, 9 and 12, 10, and 11. Field 10 is slightly larger than the others, but it has several acres of wildlife land and of nonproductive drainage way, which makes it effectively the same size as the others.

Since the soil in field 3 is well adapted to continuous row crops (corn, in Indiana) and the farmer needed a large supply of hay for his livestock, it was decided to use a rotation of row crop-small grain-meadow-meadow on fields 8, 9 and 12, 10, and 11. This represents the maximum use to which the class III land in field 10 should be placed. The class II land in field 11 will not be used to its maximum capacity; however, this rotation would constitute a more intensive use of the class IV land than is permissible. Therefore the class IV land in field 11 will be left in a third year of meadow when the remainder of the field is put in row crop. Class I land in fields 8 and 9 and 12 will not be used to its maximum, because it lies in an irregular pattern with class III land. No practical method exists by which class III land on short, irregular slopes can be sufficiently protected to permit it to be farmed in continuous row crop. The class II and class III land in fields 8 and 9 and 12 can be safely farmed in the four-year rotation by use of diversion terraces and contour tillage.

Some small areas of potentially tillable land were left in permanent pasture because of specific difficulties in using or developing them. The class II land in the north part of field 5 was left as pasture because it is relatively inaccessible to farm machinery. Class I and class

II land in field 6 was left as permanent pasture. The class I area is very small and could not be farmed as a separate unit economically. The class II land needed drainage, and the only outlet available was across the neighbor's field to the west. It was impossible to work out a satisfactory arrangement for accomplishing this. Therefore the only alternative was to develop the class II land into high-quality pasture.

### *Crop Utilization*

Since the use of sod crops is one of the best methods of erosion control, a farm on which a soil-conservation plan is used frequently produces a rather large amount of forage. This may bring about a problem of utilization. Since fertilization is normally greatly increased with the adoption of a conservation plan, the grain crops also give a higher per-acre yield, compensating partially for the reduced area planted to grain. Many farmers have found that less grain and concentrates are needed to balance high-quality forage than are required to balance a ration of poor pasture and hay. In some cases purchase of greater quantities of concentrates may be needed or a change in the type and numbers of livestock fed may be advisable.

## SPECIAL CONSERVATION MEASURES

Mulch tillage, terracing, contouring, strip cropping, and subsoiling are some of the methods designed to conserve the soil while still using the land intensively for the production of high-value grain and row crops. It is the task of the farm planner to balance the use of such methods with the change of land use, so that the result will be maximum production plus full maintenance of productivity. In addition to this, such soil-conservation measures as sod waterways, gully repair, barnyard protection, wind-erosion control, and other practices may be needed on the individual farm. They will have to be included in the over-all farm plan so that it can be estimated when these improvements can be made and what materials and expenditures will be needed.

### *Coordination of Various Soil-conservation Methods*

When planning soil conservation for a farm, it is essential that the methods used are well integrated in order to achieve the greatest results with the minimum of effort. Land-use adjustments, proper tillage, fertilization, and water management are usually the basis for specific soil-conservation techniques and therefore will generally have to precede them. Where vegetative control is possible and can accomplish soil conservation at the same or at lower cost than mechanical control, it should be used first. Grassed waterways are probably the most important conservation method after the first four items of good agriculture, land use, tillage, fertilization, and water management. Under any conditions the farm has to be considered an economic unit, and the land-use distribution has to be based on the need for the various crops as well as on the land-use capability. This means that a proportion between cropland, pasture land, and woodland has to be found that is best suited to maintain a successful farming enterprise under the given conditions. Frequently it appears desirable to have a large part of the cropland in sod crops in order to keep soil losses within permissible limits, yet there may not be an economic need for such a large amount of forage. In such a case other soil-conservation measures, for instance, mulch tillage, contouring, and terracing, have to be stressed, so that it is possible to grow more row crops and yet maintain the soil. Placing field boundaries and fences on the contour is an excellent means of saving soil without reducing the productive area of a farm.

Recreation for the farmer and his family and wildlife refuges should be integral parts of every farm. Odd areas that do not fit well into the individual fields may serve for these purposes even though their land-use capability may be of a higher order.

### *Integration of Soil Conservation and Economics*

Once the entire soil-conservation plan has been established on paper, it is necessary to calculate anticipated yields, expenditures, labor requirements, and income for a period of several years before actually beginning with the change-over. While conservation of soil is a necessity for the maintenance of soil productivity, any method devised to accomplish this successfully will have to be so designed that it can be financed without difficulty. Where the farmer possesses ample means, this will be a small problem, but where the financing has to be done from current income, a slower change-over may be necessary. In no case will a conservation plan be acceptable where the long-time income will be less than it was before its adoption. In doing this calculation, however, the value of the soil loss will have to be included in order to arrive at an equitable comparison.

One of the factors of conservation farm planning that deserves close attention is the economical use of labor. Methods of simplifying each task on the farm have to be considered. Diversification of crops and increased livestock numbers and consequently a balance of labor requirement throughout the year are frequently the result of adoption of a soil-conservation program.

## CONSERVATION PLANNING FOR LARGER AREAS

While the unit of land management is the farm, several soil-conservation tasks go beyond the boundaries set by property lines and have to be attacked by community effort. Where the distance to a natural drainage outlet is large, a ditch may pass through, and serve, several farms. Therefore it will have to be planned and paid for by several landowners. Stream regulation, especially in the case of rivers, requires the work of larger groups, perhaps even as large as state organizations.

The soil conservationist faces a special task in the consolidation of scattered farm holdings as they exist in Western Europe and elsewhere. The transfer of land from father to children in those areas has meant fragmentation of the farms, splitting fields into many narrow strips. This results in a large number of widely scattered plots that make farming difficult and uneconomical. The process of redistribution of ownership to consolidate the farm units provides an opportunity to arrange farm and field boundaries in such a way that the erosion hazards are minimized.

Another example of soil-conservation planning for areas larger than a farm is the so-called "zoning." This is the setting up of local ordinances determining the use that can be made of the land. This can refer to the assigning of certain areas for agricultural, forestry, industrial, commercial, residential, or recreational purposes, or it may be more specific and prescribe the crops and the management of the land.

# 9

## *The Future of Land Management*

Land management has changed throughout the history of agriculture, and it is different in the various parts of the world. During the last one hundred years the changes have been particularly rapid. It is interesting to attempt to take a look into the future.

The increase in population raises the value of the soil. More land is continuously used for residences, roads, industries, and other non-agricultural purposes. To make the remaining land produce enough to feed and clothe all people, more effort has to be spent. It is quite obvious that a farmer will do all he can to conserve a soil into which he has put much labor and expense. We must assume, therefore, that in the future necessity will force us to place an ever-increasing emphasis on soil conservation.

The attitude toward the land is as varied as the people themselves. Nations with restricted areas for food production have frequently developed a high regard for the land. On the other hand, where abundant land can be had for little effort, not much consideration is given to it. It is obvious that the ever-decreasing area available to the individual will cause people to become more land conscious and conservation-minded.

The recognition of the importance of soil for the survival of the human race has resulted in extensive conservation education in the United States and elsewhere. The conservation attitude, together with technical conservation knowledge, is taught from grade school through university; it forms the central theme of conservation field days, arbor days, conservation-education camps and workshops, and of many articles in magazines and newspapers. Today, bankers recognize the value of soil conservation for the safety of farm loans, and the medical profession understands the influence of soil management on the nutritional value of the food produced. Industry is aware of the influence soil conservation has on the water supply and on the wealth of the population. There is no doubt that the recognition of the correlation between soil management, economic success, and survival is going to increase further conservation teaching and result in a more active conservation attitude.

It is, of course, possible to pay the farmers to induce them to use conservation measures. The same goal might be reached by forcing them through law. It seems, however, to be the more democratic and, in the long run, the more successful approach to convince people of the economic advantages of soil conservation and of their duty to the nation, and to create a habit of conservation and make it a social misdemeanor to neglect and mistreat the soil.

In a system of free enterprise the individual farmer will practice soil conservation more extensively as he becomes aware of the economic advantages resulting from the use of such practices on his own land.

Great strides have been taken in agricultural technology during the past decades. Farmers everywhere are anxious to keep up with any developments that promise to increase yields and to improve the soil. Research in all phases of agriculture continues at an ever-increasing pace, giving promise of more efficient methods of crop and animal production and of soil conservation. New tillage methods, fertilization, and soil conditioners will help to improve soil structure and to decrease erosion rates.

The outlook for the conservation of our soils is excellent.

# *Name Index*

# Subject Index